63

11/24/2012

9.99

D0933324

WITHDRAWN

Grand Marais Public Library
104 2nd Ave. W.
PO Box 280
Grand Marais, MN
55604

# About Island Press

Since 1984, the nonprofit Island Press has been stimulating, shaping, and communicating the ideas that are essential for solving environmental problems worldwide. With more than 800 titles in print and some 40 new releases each year, we are the nation's leading publisher on environmental issues. We identify innovative thinkers and emerging trends in the environmental field. We work with world-renowned experts and authors to develop cross-disciplinary solutions to environmental challenges.

Island Press designs and implements coordinated book publication campaigns in order to communicate our critical messages in print, in person, and online using the latest technologies, programs, and the media. Our goal: to reach targeted audiences—scientists, policymakers, environmental advocates, the media, and concerned citizens—who can and will take action to protect the plants and animals that enrich our world, the ecosystems we need to survive, the water we drink, and the air we breathe.

Island Press gratefully acknowledges the support of its work by the Agua Fund, Inc., The Margaret A. Cargill Foundation, Betsy and Jesse Fink Foundation, The William and Flora Hewlett Foundation, The Kresge Foundation, The Forrest and Frances Lattner Foundation, The Andrew W. Mellon Foundation, The Curtis and Edith Munson Foundation, The Overbrook Foundation, The David and Lucile Packard Foundation, The Summit Foundation, Trust for Architectural Easements, The Winslow Foundation, and other generous donors.

The opinions expressed in this book are those of the author(s) and do not necessarily reflect the views of our donors.

Grand Marais Public Library
104 2nd Ave. W.
PO Box 280
Grand Marais, MN
55604

Society for Ecological Restoration International

**The Science and Practice of Ecological Restoration**

Editorial Board
James Aronson, editor
Karen D. Holl, associate editor
Donald A. Falk
Richard J. Hobbs
Margaret A. Palmer

*The SER International Primer on Ecological Restoration*, Science and Policy Working Group

*Wildlife Restoration: Techniques for Habitat Analysis and Animal Monitoring*, by Michael L. Morrison

*Ecological Restoration of Southwestern Ponderosa Pine Forests*, edited by Peter Friederici, Ecological Restoration Institute at Northern Arizona University

*Ex Situ Plant Conservation: Supporting Species Survival in the Wild*, edited by Edward O. Guerrant Jr., Kayri Havens, and Mike Maunder

*Great Basin Riparian Ecosystems: Ecology, Management, and Restoration*, edited by Jeanne C. Chambers and Jerry R. Miller

*Assembly Rules and Restoration Ecology: Bridging the Gap Between Theory and Practice*, edited by Vicky M. Temperton, Richard J. Hobbs, Tim Nuttle, and Stefan Halle

*The Tallgrass Restoration Handbook: For Prairies, Savannas, and Woodlands*, edited by Stephen Packard and Cornelia F. Mutel

*The Historical Ecology Handbook: A Restorationist's Guide to Reference Ecosystems*, edited by Dave Egan and Evelyn A. Howell

*Foundations of Restoration Ecology*, edited by Donald A. Falk, Margaret A. Palmer, and Joy B. Zedler

*Restoring the Pacific Northwest: The Art and Science of Ecological Restoration in Cascadia*, edited by Dean Apostol and Marcia Sinclair

*A Guide for Desert and Dryland Restoration: New Hope for Arid Lands*, by David A. Bainbridge

*Restoring Natural Capital: Science, Business, and Practice*,
edited by James Aronson, Suzanne J. Milton, and James N. Blignaut

*Old Fields: Dynamics and Restoration of Abandoned Farmland*,
edited by Viki A. Cramer and Richard J. Hobbs

*Ecological Restoration: Principles, Values, and Structure of an Emerging Profession*,
by Andre F. Clewell and James Aronson

*River Futures: An Integrative Scientific Approach to River Repair*,
edited by Gary J. Brierley and Kirstie A. Fryirs

*Large-Scale Ecosystem Restoration: Five Case Studies from the United States*,
edited by Mary Doyle and Cynthia A. Drew

*New Models for Ecosystem Dynamics and Restoration*,
edited by Richard J. Hobbs, and Katharine N. Suding

*Cork Oak Woodlands in Transition: Ecology, Adaptive Management, and Restoration of an Ancient Mediterranean Ecosystem*,
edited by James Aronson, João S. Pereira, and Juli G. Pausas

*Restoring Wildlife: Ecological Concepts and Practical Applications*,
by Michael L. Morrison

*Restoring Ecological Health to Your Land*,
by Steven I. Apfelbaum and Alan Haney

---

**ABOUT THE SOCIETY FOR ECOLOGICAL RESTORATION INTERNATIONAL**

"The Society for Ecological Restoration International (SER) is an international non-profit organization whose mission is to promote ecological restoration as a means to sustaining the diversity of life on Earth and reestablishing an ecologically healthy relationship between nature and culture. Since its foundation in 1988, SER has been promoting the science and practice of ecological restoration around the world through its publications, conferences, and chapters.

SER International is a rapidly growing community of restoration ecologists and ecological restoration practitioners dedicated to developing science-based restoration practices around the globe. With members in more than 48 countries and all 50 U.S. states, SER is the world's leading restoration organization.

If you wish to become a member, contact SER at 285 W. 18th Street, #1, Tucson, AZ 85701. Tel. (520) 622-5485, email: info@ser.org. www.ser.org.

The opinions expressed in this book are those of the authors and are not necessarily the same as those of SER International."

# Restoring Ecological Health to Your Land

Steven I. Apfelbaum and Alan Haney

With illustrations by Kirsten R. Vinyeta

Society for Ecological Restoration International

Grand Marais Public Library
104 2nd Ave. W.
PO Box 280
Grand Marais, MN
55604

Copyright © 2010 Steven I. Apfelbaum and Alan Haney

All rights reserved under International and Pan-American Copyright Conventions. No part of this book may be reproduced in any form or by any means without permission in writing from the publisher: Island Press, 1718 Connecticut Avenue NW, Suite 300, Washington, DC 20009.

Island Press is a trademark of The Center for Resource Economics.

Library of Congress Cataloging-in-Publication Data

Apfelbaum, Steven I., 1954–
Restoring ecological health to your land / Steven I. Apfelbaum and Alan W. Haney.
p. cm. — (The science and practice of ecological restoration)
"Society for Ecological Restoration International."
Includes bibliographical references and index.
ISBN-13: 978-1-59726-571-3 (cloth : alk. paper)
ISBN-10: 1-59726-571-3 (cloth : alk. paper)
ISBN-13: 978-1-59726-572-0 (pbk. : alk. paper)
ISBN-10: 1-59726-572-1 (pbk. : alk. paper) 1. Restoration ecology. I. Haney, Alan W. II. Society for Ecological Restoration International. III. Title.
QH541.15.R45A64 2010
639.9—dc22

2009036671

Printed on recycled, acid-free paper

Manufactured in the United States of America

10 9 8 7 6 5 4 3 2 1

The information in this book is accurate to the best of the authors' knowledge. Neither Island Press nor the authors are responsible for injuries that may occur as a result of the restoration procedures or activities described in this book.

# CONTENTS

# PREFACE

While the legal protection of land for purposes of conservation has rapidly increased in the past decade, in the United States and elsewhere, remarkably little has been done to address the health of protected lands or their ecosystems. The growing movement throughout the world to secure land with outright purchase or conservation easements by state and federal agencies, conservation organizations, and ecologically minded landowners has racked up an impressive acreage of protected lands. Most land, however, has suffered decades of abuse and is being further transformed by invasive species, rapid climate change, alteration of hydrology, and air pollution. This book is part of a series sponsored by the Society for Ecological Restoration International (SER) to encourage and guide land stewards, including owners, who wish to restore ecological health to their land. It provides a nontechnical introduction to ecological restoration for those with limited background or experience, and it complements the SER *International Primer on Ecological Restoration* (2004).

The principles and guidelines we provide are based on our personal experience with ecological restoration as well as our research throughout North America and beyond. Collectively, we have spent seventy-five years on researching some of the healthiest ecosystems on Earth and conducting restoration of unhealthy ecosystems. We have tested our methods, as well, on our own land: Steve Apfelbaum and his partner, Susan Lehnhardt, at Stone Prairie Farm, and Alan Haney at Star B Hill. While Alan's career is largely devoted to teaching and research, Steve's is focused on consulting and research. Steve's ecological consulting company, Applied Ecological Services Inc., annually works with hundreds of clients throughout the world, each project providing new insights and opportunities to evaluate restoration techniques. We, Steve and Alan, have collaborated in our teaching, research, and consulting since 1975. This book summarizes the approach we have found to be most successful in restoring ecosystems, and it emphasizes the importance of working with the natural processes on the land.

Our collaboration began with studying the effects of disturbances in the southern boreal forest in pristine wilderness to understand how ecosystems restored themselves after wildfire and wind disturbances. Later, during our long-term studies of critically endangered oak savanna ecosystems, we increasingly encountered human disturbances that impeded the repair processes, resulting in loss of native diversity, rapid runoff and soil erosion, and reduced water quality and productivity. We spent countless hours in conversation around campfires exploring how we might assist damaged ecosystems regain ecological health.

As we witness increasing ecosystem deterioration, we have become more compelled to promote ecological restoration. The threats to natural ecosystems are growing steadily and have reached the scale of our entire planet. Ecological restoration is no longer an option; it is essential to our physical and spiritual survival and to the lives of other species with which we share the Earth.

# ACKNOWLEDGMENTS

Our understanding of ecosystems is the result of our own research,projects, and countless interactions with others: researchers, ecological restoration practitioners, and observers of nature. These friends and colleagues are landowners, property managers, students, and even casual acquaintances with whom we have discussed observations and insights. It is futile to attempt to list all who have contributed directly, much less indirectly, to this book, but many must be noted because of their special contribution.

Hundreds of students and colleagues have assisted with our studies and contributed to the discussions as we began to make sense of observed patterns. We especially want to recognize students from Warren Wilson College and the University of Wisconsin–Stevens Point whose help in the field and persistent questions led us to insights and understanding that otherwise might have been missed. We acknowledge Miron (Bud) Heinselmann and Ed Lindquist for their encouragement in our early work in the southern boreal forest. We began our studies of oak savannas in the Midwest twenty-five years ago, with support from many organizations, notably the Division of Nature Preserves of the Indiana Department of Natural Resources, the U.S. Environmental Protection Agency, the Illinois Nongame Research Funds, and the Sand County Foundation. It was this research, especially, that helped us gain many insights into ecosystem responses to restoration treatments.

By far, the most direct and important contributors to our knowledge of ecological restoration are colleagues at Applied Ecological Services Inc. (AES). AES projects across North America and around the world each added additional insight and understanding, captured in thousands of reports, most of which were initially prepared by others. Much in this book represents a summary of that knowledge. Several AES staff members contributed directly to the preparation of this book through reviews, assistance with figures, and preparation of the manuscript. We wish to specifically mention Jacob Blue, Kim Chapman, Jason Carlson, Jill Enz, Matt Kocourek, John Larson, Doug Mensing, Lynnette Nelson, and Mark O'Leary who reviewed early drafts of the manuscript. Recognition also must go to Susan Lehnhardt, who not only reviewed early drafts of the book but was involved in countless discussions leading to our insights, and who worked elbow to elbow with Steve in the restoration of Stone Prairie Farm. Andrew Strassman also critiqued early drafts of some chapters. We extend special thanks to Lora Hagen who provided detailed reviews and patiently offered editorial guidance. Finally, thishis book would not have been completed without the encouragement and guidance of Barbara Dean, executive editor at Island Press.

# INTRODUCTION

The aim of ecological restoration is to restore ecological processes that have been damaged or lost. Ecological processes, such as succession, soil development and maintenance, and pollination, all depend on native diversity—the organisms that collectively make up the natural communities within each ecosystem. In restoration, we often focus on restoring conditions that permit reestablishment of the missing or diminished species critical to the ecosystem. Throughout, however, the focus should remain on ecosystem processes.

Humans more often than not have run roughshod over the land, using or taking what we want, with little understanding or regard for how our actions fundamentally have altered ecological processes. We have moved species from one region to another, altered the atmosphere, and released substances into the air and water that now impact every ecosystem on Earth. Some of these changes have happened quickly, but more have occurred slowly, with insidious shifts in ecological balances. As a result, most changes have gone unrecognized as long-term threats to the quality of human life.

Perhaps the growing concern over climate change, more than other environmental shifts, has opened our eyes to the realization that we live in a finite ecosystem, with definite limits to the abuse it can sustain while continuing to provide a quality environment for us. It is becoming clearer that all species and all ecological processes are interconnected, although the connections are often not obvious. We can no longer ignore the cumulative effects of invasive species, inappropriate land uses, and pollution in all its varied forms, the consequences of which are unhealthy ecosystems that are less able to meet our aesthetic as well as pragmatic needs.

Teaching ecological restoration is a bit like teaching parents how to nurture their children. While some of parenting and restoration is learning to be aware and using trial and error, we do not discount the importance of knowledge and experience. In this book, we draw from our extensive experience and that of others, as well as basic knowledge of ecology, to develop a systematic and structured approach that will increase your chance for success in restoration projects, whether you are just getting started or have years of experience.

This book, however, can cover only the basics of how to restore an ecosystem, much as a medical book can provide only the basics of how to cure human illness. Indeed, ecosystems are far more complex than humans, with thousands of species and nearly infinite relationships among them and their environments. The exceedingly complex knowledge of ecosystem form and function fills libraries, where individual books can take the reader into the intricacies of the many details. However, just as a mother can provide for the needs of an infant without a formal education, or a beginning doctor can put a patient back on the road to better health, this book can guide the beginning steward who is motivated to restore an ecosystem. Ecosystem restoration is neither quick nor easy, but it is possible, given opportunity, patience, and determination.

The primary focus of this book is on *how* to do ecological restoration, but understanding *why* is equally important. Thus, we have three objectives: to point out the benefits of healthy ecosystems, to share the joy and challenges of restoration, and to provide practical concepts and techniques for restoring ecosystems. You likely would not be reading this if you were not interested in learning and doing ecological restoration. Our goal is to give you enough knowledge to recognize and understand restoration needs, and guide you in gathering sufficient information to make informed decisions and organize and implement a restoration strategy.

Recognizing whether investment in restoration should be attempted is not always easy. Knowing whether or not you can or should try to help nature is sometimes not obvious. Perhaps, like Steve's, your property has been farmed and you want to restore it, insofar as possible, to more natural conditions. Recognizing that need is relatively easy. Perhaps, on the other hand, you own a wooded property that appears to have been abused by careless logging. The first three steps of our ten-step process, described in chapters 2 and 3, will help you understand whether or not you can or should be able to jump-start the process of restoration. Do not be discouraged if it turns out that the best thing you can do is let nature alone. You will still have accomplished a great deal in establishing a close relationship with your land, by learning its history and becoming familiar with its ecosystems and the ongoing ecological processes.

Most books on ecological restoration require at least a basic knowledge or some formal education in ecology. This book was written for readers with minimal experience or ecological training. A basic glossary is provided to assist in understanding some of the more technical terms that are unavoidable, or that you will encounter in gathering the information needed for your projects. Much of what is known about the actual practices of restoring ecosystems is based on experiential learning, gained from field projects, and we used that experience to lay out the restoration framework. We are confident that following the steps we recommend will enable you to avoid some of the difficult lessons we learned through mistakes. The framework also will assist in organizing your own observations, technical knowledge, and experiences, leading to increased likelihood of success.

For those who wish to go further, we provide a list of references and additional resources, including Web sites. We have avoided citing technical literature and have chosen sources generally available in public libraries. For those who complete this book and want more, we recommend a wide range of advanced books on ecological restoration that can be found through the Island Press Web site (www.islandpress.com/bookstore/index.php). You might also want to browse the Society for Ecological Restoration Web site (www.ser.org/) for additional information.

This book is organized into two parts. Part 1 focuses on restoration principles, concepts, and techniques. While we have used the restoration of a southern Wisconsin landscape as an illustration in this first section, the principles and concepts, as well as most of the techniques, apply more broadly to all ecosystems.

In part 2, we examine the challenges and recommended techniques for restoring ecosystems in biomes throughout North America. You will find the same approach in each type of ecosystem we discuss, although there are some variations and unique aspects to some—streams for example. In every example, you will find the same sequence of discovery, planning, and implementation.

We have written this book with a forthcoming companion workbook in mind, which will make available standard restoration planning tools, including worksheets for determining equipment and plant material needs, schedule-planning aids, monitoring support tools, and project-costing spreadsheets. These forms and tools will facilitate your collecting and processing the information needed to make good decisions about your restoration projects.

# PART I

# Principles of Ecological Restoration

Principles of ecological restoration, the focus of part 1, apply to every kind of ecosystem. We begin the book with an overview of the process (chapter 1), using the restoration of Steve Apfelbaum's Stone Prairie Farm as a case study. This run-down dairy farm was restored to a rich prairie-wetland-savanna landscape by Steve and his partner, Susan Lehnhardt.

In chapter 2, we explore the meaning and recognition of ecosystem health and dysfunction. We then dissect the process of ecosystem restoration, breaking it down into ten steps, continuing with Stone Prairie Farm as an illustration. The following three chapters provide the details for each step of restoration, beginning with chapter 3, where we describe the first eight steps of the ten-step process, including development of a restoration plan. In chapter 4, we expand on the principles that guide ecological restoration and describe in some detail the most commonly used practices that are employed during implementation. Chapter 5 expands on the importance of good goals and objectives and a monitoring program to evaluate progress toward them. We urge that monitoring be used to regularly review the restoration project, and that you share results with stakeholders and others who may be interested in the process.

Ecological restoration is an iterative process. You probably will be surprised, as we have been, at how much each project can teach you about the ecosystems you are addressing. This learning is a delight to those of us interested in nature, but it also is an important teaching opportunity for kids, neighbors, and family members, among others. We encourage you to involve as many stakeholders in your projects as possible, and share the work as well as the joy that comes from restoring health to the land.

## Chapter 1

# Connecting with the Land: The Story of Stone Prairie Farm

*I only went out for a walk and finally concluded to stay out till sundown, for going out, I found, was really going in.*

John Muir

Ecological restoration is an affair with nature. To understand an ecosystem or landscape you must become intimately involved. In doing so, you will discover, as John Muir suggested, that you are drawn emotionally and even spiritually to it, much as a person becomes attached to a mate. As with people, each ecosystem is unique. Your challenge is to discover those unique qualities, and seek ways to allow the natural characteristics of the ecosystem to be restored to a healthy condition, ultimately to stand on its own with minimal support. In this chapter we describe the discovery process and restoration of a damaged landscape in southern Wisconsin where Steve Apfelbaum, and later, his partner, Susan, developed and refined a systematic approach to ecological restoration that has been successfully applied to hundreds of ecosystems around the world. In the following four chapters, we describe and explain the process of ecological restoration in step-by-step detail, so don't despair at not getting enough information in this first chapter. Our aim here is to describe the overall process, through Steve's experience.

Undoubtedly, you opened this book because you are interested in nature, and more specifically, you are interested in restoring degraded or damaged land. You may already be intimately familiar with your own ecosystem or landscape. It may be thousands of acres of overgrazed rangeland or a backyard. The restoration process we describe is applicable to any landscape or ecosystem, even those that have been destroyed and that must be literally reconstructed from the ground up. While we can not promise that you always will be completely successful, or that the job will be easy, we can promise that if you roll up your sleeves and involve yourself in the process, you will learn more about nature and your land than you can imagine.

Perhaps you own property, as Steve did, where you wish to restore nature, or maybe you are a manager of property and have been asked to do ecological restoration. Whether you are a property owner, a manager, or simply someone who enjoys nature, you are beginning an adventure. Perseverance will be important. In the beginning, everything will be new and exciting, but like any good relationship, ecological restoration involves determination and work. Most important is understanding a process and finding the path to work with the land.

When Steve began restoring the ecosystems hidden beneath the façade of an overworked dairy farm in Wisconsin, described in *Nature's Second Chance,*[1] he was relatively inexperienced. The art and science of ecological restoration were in their infancy. Steve had formal training in ecology, plant taxonomy, soils, and related subjects, and several years' experience working as an ecological consultant, but the systematic steps for ecological restoration that we describe in this book were largely learned by trial and error. The validity of the process can be seen in successful restoration of phosphate mines in Florida, riparian wetlands in Louisiana, coastal dunes along the Atlantic coast, forests in North Carolina, and deserts in Nevada. We use Steve's experience on his land to introduce the process, then discuss later how the process works in other kinds of ecosystems.

## Exploring the Landscape

Steve first saw the land on a hot July day in 1981. As he got out of the car, he could see forests and savannas a short distance to the north. In nearly all other directions, farm fields dominated the rolling landscape, but he could see dark prairie soil between rows of corn and soybeans. Along the roadside and in the fencerows was a variety of prairie plants, remnants of the past when prairie swept across the hills from horizon to horizon, broken by scattered patches of savanna. Just southeast of the rundown farmhouse was a colorful hill, with patches of yellows and purples that could only be yellow coneflower, silphiums, pale purple coneflower and, perhaps, bee balm. The house could wait. Steve had to explore the land.

For several hours, Steve wandered across the old farm. From the window of the hayloft of the old barn, he looked beyond rows of corn growing on much of the farm and neighboring land. He could see a stream course outlined by old willows and box elder trees. Trees and shrubs also marked perimeter fence lines. The stream and fence lines continued down the valley, across neighboring farms, slicing through thick fields of corn.

For Steve, there was an organic linkage between the old house and barns, and the landscape on which they sat, the latter being more important. Buildings are more easily restored than ecosystems, and it was the landscape on which Steve and later his partner, Susan, would focus their energy. Although the landscape of the old farm was only eighty acres, it contained many different ecosystems in various stages of dysfunction, some disturbed, others damaged, and some even destroyed, replaced by nonnative weeds.

On his first visit, Steve was especially intrigued by the stream that wound through a pasture along the lower margins of cornfields. Although wetland vegetation was largely absent, the stream was bordered by black "mucky" soils that could only have been developed under former wetlands, where accumulating plant matter decomposes slowly. Wading through deep patches of nettles, thistles, and ironweeds, he stumbled onto a cattle trail and followed it to the creek. There he found green, bubbling ooze over the surface of tepid water, and bare, mud-caked banks on either side. Dairy cattle were standing in the creek, their tails actively swishing the flies drawn to the stink. Water, whether in lakes or streams, is often the best indication of the ecological health of the landscape in which it occurs. When landscapes are abused, soil, nutrients, and contaminants move, most often ending up in a waterway.

The condition of the creek on Steve's property was a result of several environmental insults, called *stressors*, some coming from his farm, others from neighboring farms upstream. Healthy ecosystems retain soil, as well as water and nutrients. When ecosystems are damaged, in this case by intensive farming practices, more water runs off and moves more quickly, with resulting soil erosion and loss of nutrients. The water in Steve's stream was being fertilized by cattle, as well as by erosion and nutrients being lost from his and neighboring croplands.

In water that is low in nutrients, algae grow slowly and are easily kept in check by fish, snails, and other invertebrates that feed on algae. Some nutrients, of course, are essential for the algae to grow and for the stream to be productive. In a healthy ecosystem, the nutrients that enter the stream are nearly equal to the loss of nutrients through animals that visit the stream to feed on the snails, fish, and invertebrates. A raccoon or a barred owl patrolling a stream may pick off a frog or snake here and there. The nutrients from the water, absorbed by the algae, converted to insects, then to frog tissue, return with the predator to some upland site where they are released as urine or feces. No doubt many predators were visiting the stream and removing prey, but the influx of nutrients from the farms greatly exceeded the rate at which the nutrients were being removed, leading to excess growth of algae.

The problem did not stop there. The bubbling that Steve noticed through the dense slime of algae was certainly methane gas, indicating anaerobic decomposition, and anoxic conditions in the water. As sun shines on a shallow, exposed stream, the stream warms, and respiration of both plants and animals increases. Oxygen produced through photosynthesis by the algae is quickly absorbed from the water by all the respiring organisms, and any excess is lost to the atmosphere. When the sun goes down, respiration continues, but photosynthesis stops. Consequently, at night, dissolved oxygen levels in the water soon drop below the level needed by virtually all air-breathing animals. Diversity of consumers decreases, and with that, the ability of the stream ecosystem to consume the excess algae, which, as it dies, is anaerobically decomposed by a host of fungi and bacteria. A vicious cycle is established.

Troubled by what he found, Steve followed the stream toward its headwaters. Upstream, above the pasture, springs seeped from the banks, crystal clear and cool. There, Steve found a variety of native wetland plants—turtle head, blue vervain, several sedges, golden-glow sunflower, mountain mint—that indicated wetland diversity was still present although excluded from the pastured stream banks by grazing activity. In the water, he found a rich variety of invertebrates. This was a tremendously important discovery. Ecosystems can be restored only if much of their natural diversity can be returned. Each species plays a role in the functioning of a healthy ecosystem. It is the interaction of species through these functions—predator-prey, plant-herbivore, disease-host, competition, parasite-host relations—that defines an ecological community and provides it with healthy characteristics, such as resiliency and stability. Overlap in these functions creates duplication such that the ecosystem will continue to function even when many species are lost. However, when too much diversity is lost, the ecosystem is less capable of responding to disturbance or climatic fluctuations. Only when most of the species are present and healthy and able to interact freely with other species will the ecological community regain higher levels of health. Thus, much of your work in restoring your ecosystem will be directed toward restoration of the native species that once were there.

Ecological *resiliency* is defined as the ability of an ecosystem to recover following perturbation. Both the rate and degree of recovery reflect the health of the ecosystem. Steve's creek had lost the ability to respond to the surge of nutrients it was receiving because it had been overwhelmed by the influx of nutrients, and too many species had been eliminated. The balance of nature in the stream had been disrupted by excess nutrients in the water. As species are lost, the ecosystem becomes less and less able to respond to the perturbation, and resiliency is lost.

Steve's joy at finding the diversity in the wetland near the headwaters of the stream was soon dampened when he found more problems. The rows of corn ran up the slopes, and the eroded soils ran down. Heavy deposits of topsoil had accumulated along the lower margins of the cornfield, just above the stream. The next storm would flush another surge into the creek. To make matters worse, nonnative reed canary grass was invading the patches of remaining native plants between the field and the stream. Whereas native species are well adapted to residual soil conditions, the eroded soil that accumulated near the creek was deep and fertile, favoring invasive species, in this case, reed canary grass. There also were nonnative trees, European fragile willow and Russian olive, along the stream. As the stream banks eroded, a result of intense runoff from the cornfields and bank disturbance by the cattle, the willows collapsed, pulling large chunks of bank with them. To add insult to injury, there were scattered old coils of barbed wire, spent farm equipment, and trash dumped indiscriminately along the stream. Steve began to realize the total landscape had been abused and neglected, not from malice, but from ignorance and, perhaps, some necessity stemming from marginal profitability.

## Developing a Plan

Steve felt overwhelmed but took solace in knowing that the farmer had a two-year lease during which a restoration plan could be developed. Perhaps it was fortuitous that Steve was forced to bide his time. The landscape had been damaged by agricultural practices over nearly 150 years, and restoration would take at least a few decades. Jumping in too quickly could lead to costly mistakes. Many temperate ecosystems, even damaged ones, are remarkably resilient, however. As long as native species can be restored, ecological functions will recover, but mistakes generally mean lost time or money. Taking time to study the land, gather information, consult with experts, and test hypotheses is the best approach in every instance. By necessity, Steve proceeded slowly, allowing time to sort through the issues, define the needs, set priorities, and determine where best to begin.

Steve also suffered two deficiencies: experience and money. While he had a good formal education, book and field knowledge are two very different things when it comes to application. Restoration ecology is still emerging as a formal body of knowledge.[2] Steve didn't even know how to start the two old tractors in the broken-down shed. So, he began by talking with his neighbors—farmers who had similar equipment and years of experience squeezing a living out of the land. He not only learned how to start and drive his tractors, he also learned much about the history of the area around his farm. Being short on money, Steve fell back on a necessary strong suit when doing restoration: ingenuity.

Collecting information and thinking about ecological changes that had occurred at Stone Prairie Farm, and the things that were needed to begin restoring the land, Steve's restoration plan began to crystallize into a ten-step process, described in detail in the following chapter. Steve felt he had no choice in what to do first. It was one thing to have erosion in the cornfields, but Steve especially wanted to stabilize soil along the road, within easy view of his neighbors. Converting manicured, clean-looking cornfields to weedy-looking, native-plant communities is a transition that might be understood and appreciated by ecologists, but likely not by neighboring farmers. There was a double standard—erosion is tolerable to farmers working the land, but it would be objectionable in the restoration that local farmers initially would see as a waste of land.

With only a few acres near the road not under lease, Steve prepared part of the area for a garden but was unsure how to convert the remaining acres into prairie. A neighboring farmer told him how to prepare the ground for oats. This also gave Steve a place to experiment. He collected native prairie seeds that autumn and worked them into the soil with a hoe before the oats were planted. The following spring, the green field of oats sent the right message. By late summer, he and Susan could find small prairie seedlings scattered through the oat straw.

## Getting Started

Not being able to start work on the rest of the farm for another year, Steve used some of the prairie seeds to develop patches around the house. From these plants they later collected some of the seed used to plant the larger fields. After three growing seasons, they had enough seed to begin converting a few acres each year, improving restoration techniques as they learned.

When the farming lease expired, Steve turned his attention to uplands and drainageways that were tributaries to the stream. Where the storm water runoff from neighboring farms needed to be controlled and cleaned, he planted perennial prairie vegetation to serve as a buffer. If the runoff couldn't be prevented, it could at least be intercepted before it got to the stream. This buffering strategy would make it possible to restore the stream without having to eliminate the sources of soil and nutrients. After the cattle were removed, Steve discovered a flush of native plants in the former pasture and along the stream course. Most areas along the stream, even some of the areas with eroded banks, self-stabilized. Where the banks did not stabilize, he had a local earth-moving contractor grade steeper banks and remove the fallen, invasive willow trees. He then seeded the banks with native species and a cover crop to stabilize the soil until those slower-growing native perennials could get established. Whenever possible, extended family and neighbors were invited to participate.

Steve began the process of stabilizing soil in former cornfields with cover-crop plantings of barley, and he used the grain to barter with neighboring farmers for needed services. In exchange for barley, farmers helped to remove fences and apply herbicide to start conversion of hay-fields to prairie. In exchange for straw, farmers also mowed the stubble to provide prairie plants with needed sunlight.

The "horse-trading" accomplished two other things. It gave Steve an opportunity to explain to his neighbors what he was doing, and why. So, in addition to getting barley and straw, neighboring farmers got some lessons in ecology and restoration. Second, their involvement in the restoration process gave neighbors some "ownership."

Ecological restoration cannot be done in isolation. What you restore is always part of a bigger system. An acre is part of an eighty-acre farm, which is part of a neighborhood, and neighborhoods are part of watersheds, and so on. Barred owls feeding in the Stone Prairie creek nested in a woods a mile to the north. The stream that crossed Steve's land also crossed farms of his neighbors. The neighbor immediately downstream, who used the stream to water livestock, noticed that the water coming from Stone Prairie Farm was clearer and cooler. Bobolinks that began returning to restored prairie also belong to South America and grassy fields between.

## Scaling Up

As prairie became established, Steve began prescribed burning. For this he needed help. Safely managing a prescribed fire requires experience, knowledge, permits in most areas, and often additional hands and equipment. Beer and spaghetti usually were enough to enlist neighbors, colleagues, and friends, but Steve also needed to line up some experienced help. Like the annual spring run of northern pike up the local streams, burning Stone Prairie Farm became a neighborhood tradition, with more fun and less excitement as he gained experience.

Through an annual process of sowing seeds collected from the garden plots and from neighborhood prairie patches, the restoration area expanded each year. Seeds collected from dry prairies in the area were planted only where dry prairie historically would have grown. Wetland species were introduced to appropriate soils that once harbored wet prairies and riparian wetlands. A rich diversity of native species was present within a few years. As he could afford it, this diversity was augmented by purchase of seeds or plants of local genetic sources that were missing but that research indicated had once been present in that landscape. In all instances, species were planted in the appropriate soil types and moisture regimes. Over time, the less-common species spread to other appropriate settings in the property. Now, years later, Steve and Susan have found that most of the species that were seeded, including many that initially seemed not to become established, eventually showed up. They learned that, with more patience, they could have saved money and hours of hand planting by giving nature more time to let seeds germinate and develop.

Patience is an important virtue Steve did not initially factor into his ambitious restoration plans. As with farming, establishing an ecosystem relies on timely rainfall and favorable growing conditions. Natural ecosystems, however, require years to become reestablished, unlike the monoculture or simple polycultures of farm crops, pampered by cultivation, fertilizers and pesticides, or irrigation. Even with careful introduction of several dozen species, only a tiny fraction of the natural diversity of an ecosystem will be represented. The thousands of associated species, from bacteria to birds and mammals, may require decades to find their way back and adjust their populations to the conditions that are present. The more one can establish connectivity with remnant ecosystems where some of the diversity still resides, the quicker the ecosystem can be restored, but patience is essential.

## Learning to Work with Nature

Steve initially tried to cut and herbicide every invading tree and shrub that showed up where he planned to restore prairies. If the planning map he and Susan had drawn said it was to be prairie, Steve was determined to make it prairie. Eventually, after cutting trees from the same area again and again, Susan urged him to rethink the plan. He decided to let the black cherry and box elder grow. Now, in places where trees are prone to grow, savanna restoration is well advanced. The trees formed thickets and shaded the underlying herbaceous weedy species that persisted after farming ceased. Seeds of native grasses, sedges, and wildflowers found in local savannas were hand-broadcast into the shade of the trees. With reduced competition from residual agricultural weeds and prairie plants, which need nearly full sunlight, savanna species have flourished.

## Principles Learned from the Stone Prairie Experience

Four important restoration principles emerged from the Stone Prairie Farm experience. First, *pay attention to what nature wants and work with her.* Otherwise, you are swimming upstream and your efforts will be greater and successes fewer.

Second, *don't insist on recreating the past.* We often have only a vague idea, at best, of what the past was, and more often than not, conditions have changed. Although from historic documents you may have developed a perfectly logical restoration plan, you can waste a lot of time and energy trying to force your plan onto the landscape. Instead, pay more attention to the first principle, and be flexible.

Third, *be patient.* In only five or six years, Steve and Susan could cast their view across what superficially looked like large expanses of prairie, but closer inspection would reveal a pittance of the total diversity that would eventually follow and was needed for the prairie ecosystem to become fully functional.

Fourth, *when in doubt, follow the first principle.* Steve and Susan also learned that the healthiest fragments of the remnant ecosystem were the quickest to respond to restoration. The more intact the ecosystem, the higher the recuperative potential, or resiliency. It frequently makes sense to start restoration with the most intact pieces for several reasons. First, these areas provide the quickest return on invested time and money. Second, these healthier areas are refugia for species that can then be introduced to other areas. Finally, some quick success goes a long way toward appeasing impatience and building confidence. Steve and Susan had quick success with some of the wetland areas along their creek, and the aquatic system responded rapidly. That is often the case, because streams are open systems. The excess nutrients are literally flushed away soon after the source is brought under control. Upland sites with damaged soils, in contrast, might require decades of patience.

The stream was a good example of where some minor modifications result in quick responses. By restoring the historic stream channel, two to three acres of adjacent farmland were largely restored to healthy wetland with minimal additional effort. With recovery of vegetation, blue-wing teal, great blue herons, frogs, red fox, raccoon, and many other species began frequenting the wetland. This success came about because Steve and Susan had interpreted the clues and studied historic photos and documents that clearly showed how the wetlands had been drained when the stream was channelized. Also, the muck soils contained viable seeds from native plants that had grown there thirty or forty years previously.

## Learning by Doing

Steve's formal education was a start, but he soon realized that he needed to learn to recognize unhealthy ecological communities, and plants in the seedling stages. Beyond basic soil science and hydrology, he had to learn to read subtle clues that indicated erosion, sedimentation, and anaerobic subsoil conditions. Formal training is helpful, but nearly *every* beginner will need to spend time in the field with people who have experience in reading nature. Only by piecing together the clues, the species of plants present, drainage patterns, soil distribution patterns, topography, and anything else that helps explain what you are seeing, can you develop best-guess interpretations and proposed restoration treatments. Trial and error, however, is a great teacher.

Each treatment at Stone Prairie became an experiment to test Steve's model of understanding. When things didn't work as expected, he went back and revised the model, trying a different approach. With each success, another piece fell into place, like working a crossword puzzle.

These functional pieces usually steer the way toward the next tasks, and increase the overall understanding of the system. The steps involve (1) carefully observing or measuring, (2) developing hypotheses and interpretations, (3) modeling how that piece of the ecosystem might respond to any given treatment, (4) implementing the treatment, (5) following up with observation or measurement, (6) revising or refining the hypothesis or interpretation and model, and so on.[3] This iterative process never ends, although as one gains more and more knowledge, it becomes a shorter process, with less testing.

## The Necessity of Stewardship

It soon became clear to Steve and Susan that they would never finish restoring Stone Prairie. With each passing year, they experience the joyful return of life, but also they are reminded that much is still missing. Because Stone Prairie Farm is no longer part of a vast natural landscape that once stretched across the land, it will always remain as an isolated fragment, an island in the middle of cornfields, pastures, roads, and towns. Partly because not all species can be restored, and partly because the farm is a fragment, there is a strong ongoing need for maintenance. If the farm were connected to large, fully functional, healthy ecosystems, maintenance would not be necessary, at least in theory. However, neighboring farms continue to release nutrient-enriched soil, nonnative weed seeds, and farm chemicals onto Stone Prairie Farm. Even if all adjoining areas from which wind and water flow could be restored, some stressors would persist. Flocks of migratory robins still would defecate European buckthorn, tartarian honeysuckle, and multiflora rose seeds each spring, and the wind would carry Canada thistle seeds to patches of soil freshly exposed by woodchuck burrowing. A vigilant and focused effort of stewardship is as important as the restoration process. In fact, it is an integral part of the process.

Burning, spot-spraying and stump or root treatment of invasive shrubs, pulling up garlic mustard, and wicking herbicide onto the invading reed canary grass are part of the annual maintenance regime. In the winter, Steve and Susan mark newly invaded shrubs with ribbons. They cut the stems and herbicide the stumps a week later. They check again in the spring and midsummer to ensure success. Once cleared of invasives, these areas are then hand seeded with appropriate species for each location, after the annual spring burn.

Steve and Susan also mow trails and fire breaks annually for surveillance and for ease in walking the land. Every walk is a monitoring operation, as they are always on the lookout for threats. This is no different from farmers walking or driving their fields, foresters their woods, or ranchers their rangelands, except Steve and Susan are watching for invasion of many of the species the others might hope to encourage. At this stage in the recovery, however, most surprises are of additional native species that have found a home at Stone Prairie Farm.

## Joy of Success

The process of restoration can provide tremendous satisfaction. Finding and collecting native plant seeds is like a scavenger hunt. You may have a list of species you are especially seeking, or you may go with an open mind to see what nature can provide. It is especially gratifying to find rare species. There is great satisfaction in having grocery bags of hand-collected seeds suspended from the rafters in the old barn, awaiting spring. There is even greater pleasure in seeing the plants grow, not only as individual plants, but in the clusters or irregular groupings that develop across the landscape. Meandering across the Stone Prairie landscape, hundreds of locally obtained native plant species now form coherent plant communities, reflecting microsite conditions of soil and hydrology. This patchiness is part of the landscape diversity, and can be seen in the wildest, least-disturbed natural areas. It makes sense to put seeds or plants of species in the microsites where they are best suited, but one should not be too presumptuous about making those decisions. Let nature vote last. It will, whether you want it to or not.

One of the pleasures of ecosystem restoration follows restored vegetation. As the prairie and wetland plants became established at Stone Prairie, the animals were quick to return to the landscape. Instead of cows, Steve and Susan now watch bobolinks, bobwhite quail, and meadowlarks. Of greatest joy to Steve, the stream now runs clear and cool. Because of the cold seepage water, it harbors many rare aquatic insects and fishes that have found their way back, or have been intentionally reintroduced from neighboring streams. Where overcropped pasture once grew, there now grow lush wetlands of hairy-node sedge, vervain, swamp milkweed, indigo bush, Turk's-cap lilies, and more.

Even the hydrology has responded as infiltration has increased beneath the prairie sod. In healthy ecosystems, minimal water is lost during rain events. Instead, the vegetation intercepts the rain, releasing it through thick litter and a tangle of roots, into soil pores kept open by diverse soil organisms and growing plant roots that incorporate organic matter deeper into the mineral soil. From the creek banks, hundreds of small springs now flow. These were apparently obscured by erosion and disturbance by cattle. Frogs and toads arrive and join together each spring by the thousands—spring peepers, American toads, green frogs, gray tree frogs, western chorus frogs, leopard frogs, among others—in the creek and restored wetlands. Sleeping on their screened porch on early May nights, Steve and Susan are serenaded by the diversity of their prairie. The cacophony of harmonizing toads, competing peepers and tree frogs, and the banjo E string call of green frogs resound the joy for this newfound habitat. Steve and Susan measure the return of diversity using bird, frog, and plant surveys. They also occasionally send water from the stream and wetlands to a laboratory to get hard data on chemistry. The sounds that lull them to sleep on warm May evenings echo the results of the tests. The prairie phoenix has risen after nearly 150 years.

## Conclusions

Humility, commitment, persistence, patience, and the development of a vision based on appropriate homework and interpretation of the land are the keys to ecological restoration. There are many details to each step of the process, described in the following chapters. The most important detail, however, is developing a realistic vision. That vision begins, appropriately, by developing an intimate relationship with the land and its condition, and understanding what once occurred there. It is there where we start our adventure in ecological restoration.

Chapter 2

# Ecological Restoration: An Overview

*In every walk with nature, one receives far more than he seeks.*

John Muir

The ecological planning process results in a graphic and expository plan that serves as both a record and a road map for ecological restoration. Like a road map, a proper restoration plan identifies where you are, where you want to go, and some alternative ways of getting there. Some routes will prove to be poor choices, and other alternatives should be chosen. Some routes may get you to where you want to go but involve more time, expense, and effort than is available. A plan, like a roadmap, cannot anticipate all contingencies, but it can keep you moving toward your restoration goals, and guide you when you have to back up and try different approaches. Ultimately, the aim of restoration is to effectively and efficiently implement on-the-ground activities that lead to desired ecological effects.

Before beginning restoration, it is necessary to evaluate carefully the ecological health of land. Dysfunction in an ecosystem involves loss of some of the services or benefits that otherwise might be realized. Healthier ecosystems not only provide greater benefits, they require less effort and money for restoration or maintenance. But how do you recognize dysfunctions in an ecosystem, or conversely, what are the characteristics of ecological health? In addition to knowledge of reliable indicators of ecological health in general, you must identify and understand the conditions of a particular ecosystem. Because each ecosystem is unique, this requires getting intimately acquainted with each ecosystem on your land and comparing it to similar ecosystems in nearby protected areas. In this chapter we explore this process in some detail then provide an overview of how to translate perceptions into assumptions and working hypotheses. The assumptions and hypotheses are models of how you think the ecosystem looked and functioned before it was altered by people.

An ecosystem flows with expressions of life. Can you recall lying on your belly as a child and watching ants in your yard? Perhaps you spent hours, as we did, poking under rocks in a creek, or chasing butterflies across a meadow. Remember the fascination of spiders spinning their webs, or mud-daubbers building clay chambers for holding the paralyzed spiders they entombed in them? To understand an ecosystem, one must connect, as children easily do, to the organisms and processes.

Of course, appreciating what you see is different from understanding, or recognizing what is missing that you should be seeing. In brief, there are two ways to gain that insight: from someone who is experienced, or through research. The latter, of course, is the primary way for beginners. Research involves study of historic documents and accounts of local natural history, as well as visits to reference natural areas that are reasonably intact remnants of historic systems and discussions with older neighbors who witnessed some of the changes to the land. These will be described in detail later. For now, let's focus on recognizing the signs of ecological health.

Before beginning restoration you must decide whether you can contribute positively to the work nature is doing. Nature continually is healing damages whether caused by natural events or people. The pursuit of steps one through six described in this chapter will both strengthen your connection to your land and help you determine if your intervention is required.

## Ecosystems

Ecosystems are the organisms together with physical constituents such as air, water, and minerals that support them. Ecosystems can be viewed from three perspectives: composition, structure, and functions (fig. 2.1). *Composition* is the species that are present in the ecosystem. The organisms often are collectively referred to as an *ecological community*. It is impossible to list all species, especially in complex ecosystems, but all species are not equally important. Those that are more common, or that contribute more to key structure or functions are the critical species. Those whose presence are critical to the survival of many others are often called *keystone* species. *Structure* is the three-dimensional configuration of the ecosystem and is represented by (1) living and dead vegetation, which includes trees, shrubs, vines, forbs, grasses, evergreen species, deciduous species, bryophytes, epiphytes, lichens, snags, fallen woody debris, leaf litter; (2) mineral structure, including characteristics of the soil; and (3) landscape structure, including rocks, plant communities, streams, ephemeral ponds, and the like. The composition gives rise, to a great extent, to the

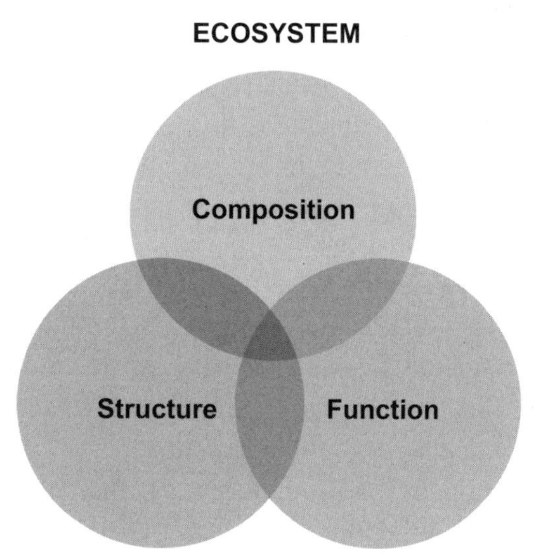


FIGURE 2.1 Ecosystems can be viewed from three perspectives: composition, structure and function.

structure and functions. *Functions* are the processes that occur in the ecosystem. All are directly or indirectly affected by the biological activities of organisms. This relationship among composition, structure, and functions is much the same in both aquatic and terrestrial ecosystems. Because the composition of the ecosystem largely determines the structure and functions, most restoration focuses on restoring or adjusting the composition. In later chapters, we will discuss some structural adjustments that are used to alter composition and functions in some ecosystems. In analyzing the health or condition of an ecosystem, composition is especially important, but structure is sometimes a key. Most functions cannot be easily seen and are best evaluated through measurements.

## *Assessing the Health of Ecosystems*

Our ability to assess ecosystem health is important for three reasons: to avoid blind continuation of ecosystem damage, to identify communities most in need of restoration, and to monitor progress during restoration. You may have a premonition that something is amiss: perhaps lack of clarity of a stream, or a patch of prairie grasses without other flowering plants, a wetland overgrown with cattails, or rangeland dominated by cheatgrass. You may have some personal experience, perhaps childhood memories of conditions that no longer exist. Perhaps an elderly neighbor recalls a species of wildflower where it no longer occurs. Most of us will have some sense of potential problems in the ecosystems with which we are familiar, and these represent beginning points for evaluating ecological health.

Evaluating healthy land has parallels in humans. How do you assess your own health? You may lack your normal energy level or suffer vague pains; you may have a fever, or a cough. With the help of a professional, you seek the cause. The doctor or nurse may measure temperature or blood pressure, run tests for blood and urine chemistry, grow cultures to identify bacteria, take x-rays, and so on. These tests, together with symptoms, may clearly indicate the cause of illness, or at least suggest a likely diagnosis and, therefore, an initial strategy for treatment. Response to treatment gives healthcare providers additional clues that can lead to refined treatments, or confirmation of correct diagnosis and treatment.

Ecosystems are incomprehensively more complex than humans, and there rarely are single cause-and-effect responses. It is far more common in ecosystems to have complex causes and effects, which may or may not be related, and which may produce many different symptoms. Prevalence of a nonnative species, for example, might be a result of altered hydrology, loss of native species, soil disturbance, elevated levels of nitrogen, or combinations of these. So as you walk the land, you need to be alert to multiple indicators of healthy or unhealthy conditions, including subtle ones. As with your own illnesses, you must take all indicators together to develop a good assessment, and you may need to consult professionals.

In contrast to assessment of human health, there is no "thermometer" for ecosystems, nor an orifice in which to insert it. While many measures are often useful to assess human health, there is no easy measure to assess health of land. Instead, we must understand trends, changes from historic conditions or patterns, and the current relationships between vegetation, soils, water, fauna, and people. Monitoring, properly done, provides us the often subtle clues for changes occurring in an ecosystem. How to set up a good monitoring system will be covered in chapter 5.

Although difficult or impossible to assess directly, ecosystem demonstrate certain characteristics:

1. Adaptability. Some of these adjustments result from the adaptations accumulated in the genetic heterogeneity within species. Others result from the diversity of species, some of which will be favored by the new condition and some of which will be disfavored. Given the myriad species present, and their effects, direct and indirect, on one another, the result is a dynamic system.
2. Resiliency. This is the capacity to return to a stable or previous condition after perturbation. Healthy ecosystems can recover from extreme distur-

bances, even clear-cutting or severe fire. The more uncommon the disturbance, or the more diversity has been compromised, the slower the response.

3. Redundancy. Many species perform similar or overlapping functions in healthy ecosystems. This contributes to efficiency, resiliency, and stability, and allows ecosystems to suffer loss of diversity without collapsing. Greater diversity generally means greater redundancy and higher resiliency.
4. Organized. Landscape patterns and communities reflect long-standing evolutionary relationships. Species, as well as genetic makeup of each species, characterize communities and reflect regional and local conditions of the landscape. Genetic makeup of a fire-adapted species, for example, will vary along fire-frequency gradients, but so will the distribution of species along the same gradient. Experienced soil scientists can map soils using the distribution of species across a landscape because species distributions reflect changes in the soil characteristics.
5. Sustainable. Each species has the innate capacity to reproduce, and the interactions among species result in communities being able to restore themselves after disturbance, the process called *succession*. A corollary to this characteristic is that similar communities will be found on similar conditions across the landscape.
6. Multidimensional. Ecological processes occur at different scales of space and time. Climate change, for example, operates at a global scale and over decades to centuries, whereas competition operates at a community scale and over days to years. Nitrogen fixation or photosynthesis operates at a microscopic scale and over microseconds to minutes. All, however, are interrelated. Likewise, we can conceptualize ecosystems at nearly any scale, from a global ecosystem to one that is microscopic. Different species also function at different scales. Microorganisms may function within cubic millimeters, whereas earthworms occupy cubic meters, deer utilize resources in many square kilometers, and black bears in hundreds of square kilometers.
7. Efficient. Ecosystems absorb and release energy efficiently . Energy is absorbed through photosynthesis, retained in biomass, and lost as heat through respiration. An ecosystem's efficiency in capturing and transforming energy vastly exceeds any artificial system.
8. Continuous. Few, if any, ecological systems are isolated. They merge one with another across landscapes and time, absorbing light energy, releasing heat energy, and exchanging gases and water with the atmosphere. The more transitions, called *ecotones*, the greater the community diversity across

the landscape, and the richer the species diversity. For this reason, it is nearly impossible to define ecosystem limits, other than the global ecosystem. Ecological processes like soil formation are also continuous, although their rates may change.

9. Functional thresholds of space or time. Some ecological functions cease if scales become too small or too short. A simple example is animals that require large home ranges. For example, timber wolf populations need 100,000 acres or more. Don't expect to restore wolves on your back forty. Likewise, evolution stops at the temporal threshold defined by the lifecycle of the species and may be imperceptible over several lifecycles.
10. Vulnerable to uncommon perturbation. Ecosystems will recover from disturbances as long as member species can survive or reinvade. The healthier the ecosystem, the less vulnerable to perturbation it is. Disturbances that are beyond the evolutionary experience of key species may permanently alter an ecosystem or render it dysfunctional beyond the ability to repair itself. In this case, active restoration is necessary.

Because the conditions of these characteristics are difficult or impossible to directly assess, we use indicators that relate, directly or indirectly, to these characteristics. The condition of each indicator must be weighed against the condition of that indicator typically found in a healthy or reference system.

## *Indicators of Healthy Ecosystems*

While hundreds of indicators for assessing ecological health might be described, only a few are so widely applicable that they justify special attention. We will elaborate later on how to evaluate these indicators. For now, we only introduce them.

- Diversity of native species appropriate to the landscape location. Diversity is the composition we referred to earlier. Diversity is not only the variety of plants and animals present, but also their organization into communities across the landscape. As environmental conditions vary from area to area, the composition of communities will reflect that variation, and with connectivity, increase the opportunities for interbreeding and migration.
- Purity of water intimately associated with the ecosystem. If there is a stream or lake in the ecosystem, it may be one of the best indicators of ecological health. High levels of dissolved nutrients and suspended particulates indicate that the watershed is "leaky" and is not conserving soil and nutrients very well. Nutri-

ents in water stimulate growth of algae and other plants that often provide an indirect clue that nutrients are leaking from the watershed. Streams may run clear except during floods, and conditions during or after storms or rapid melting of snowpacks. Examination of streams when flows are high is especially useful.

The healthiest ecosystems are frugal, or tight, meaning they retain soil and nutrients. In a seminal study in New Hampshire, a forest was cut and the area was sprayed with herbicide for three years to prevent plant growth.[1] Predictably, soil and nutrient contamination in the stream draining the area surged. When spraying was stopped, nitrate levels, the most sensitive indicator of leakiness of an ecosystem, returned to background levels within three years, as plants began recolonizing the soil. Well before the trees grew back, the release of nutrients, soil particles, and rate of storm water runoff approached the levels that occurred before the trees were removed.

- Healthy populations of native species. Genetic heterogeneity of a species cannot be maintained without large, interacting populations over broad geographic areas. When there are small, isolated populations, a species can soon lose fitness in a changing environment. Conversely, nonnative and invasive species are a sure indication of problems.
- Stable soils. The quickest way to determine if soils are stable is to examine the profile, especially noting the depth of topsoil. Such measures, however, require a baseline against which to compare. Either refer to the soil description found in the country soil survey, or check topsoil and/or organic matter in nearby, protected pockets that the soil map for the area indicates are the same soil type.
- Normal stream hydrology. Healthier ecosystems retain water better. Living organisms maintain a more porous soil with higher levels of organic matter. Plants not only intercept precipitation, their litter covers the soil and roots bind the soil, and water more easily infiltrates rather than runs off. Even water that runs off carries less soil and nutrients when the ecosystem is intact. As ecosystems are disturbed, with corresponding loss of species and vegetation, soil becomes increasingly vulnerable to erosion. Streams reflect this not only in higher levels of suspended and dissolved loads, but also in greater fluctuation in water volume, with both higher flood volume and lower volume during dry periods.
- Diverse, healthy communities. Here the focus is on the kinds of ecosystems within a landscape and the composition of each. Communities reflect the evolutionary history between species, climate, soils, and disturbance at that

location within the landscape. Look for different assemblages of plants that correspond to changes in topography or hydrology. Also note the composition within communities. One might "restore" a community of native species that is completely artificial, with no parallels in nature. Wildflower gardens are an example; unless carefully done, they require considerable maintenance. A comparison with communities in nearby reference natural areas of the same ecosystem type can suggest the types of communities and species compositions.

### *Importance of Diversity*

Among the dozens of ecological parameters that relate to ecological health, native plant and animal diversity is arguably the most important, and one of the most easily assessed. Ecological processes are the direct and indirect result of the interactions and activities of organisms. Native organisms that make up the communities in healthy ecosystems have coevolved for hundreds and thousands of generations, fine-tuning the intricate relationships among them and with their environments. A secondary indication of ecological health is the presence of nonnative species, or an overabundance of one or more native species. Most other indicators, such as stable soils and clear water, are long-term or indirect manifestations of a healthy plant and animal diversity.

Another dimension of diversity, on a longer timescale, is adaptability. Populations will shift with changing conditions. As conditions change, some species will experience greater stress, and this often is manifested by increased disease, parasitism, and mortality, or reduced reproduction and declining populations. Others will benefit and increase. Increased diversity, both in numbers of species present as well as genetic diversity within species, allows healthy ecosystems to be more responsive both in the short term and long term. For example, when chestnut blight, caused by an introduced fungus, swept through eastern deciduous forests, American chestnut was reduced in both numbers and stature, from a dominant tree to scattered shrubs. Chestnut oak and red maple increased, and the overall productivity of the forest scarcely changed.[2] Meanwhile, the chestnut trees with some resistance to the fungus continued to flower and produce offspring, the descendents of which now show evidence of becoming resistant. Of course the fungus also continues to evolve, so it is not at all clear that American chestnut will ever overcome its susceptibility, and it is quite doubtful that even if it did, it could regain its former dominance in deciduous forests, at least not in less than many millennia.

It is quite easy for an experienced naturalist to tally the kinds of common or dominant plants, insects, or birds in a terrestrial ecosystem, or invertebrates in a stream. Greater diversity in one life-form, birds for example, usually indicates greater diversity in others, for example, invertebrates. Likewise, presence of nonnatives in one life-form often will be paralleled by nonnatives in other life-forms. With experience, one can assess whether the populations are greatly different from those of similar, healthy ecosystems. For example, common reed grass may become an aggressive species in disturbed wetlands, in part, because of genes from European strains that have contaminated our native population. Its overabundance is merely the symptom that something is amiss. Nonnative species almost always have negative impacts on ecosystem functions and over the long term will result in less native diversity.

## Ten-step Process for Ecological Restoration

The ten-step process we introduce here and describe in detail in the following chapter was developed and refined over thirty years of restoration experience with hundreds of ecosystems, ranging from streams to deserts in temperate and tropical regions of the world. The process includes the initial step of assessing ecosystem health.

Restoration draws from many fields, from history to ecology. At first blush, it can seem overwhelming. The methods we describe, however, can be implemented by a lay person who is persistent and willing to seek experienced assistance when needed. A systematic approach is essential to organize the many details of information needed to make informed decisions, and to proceed in an efficient and ecologically sound manner.

The first two steps can be done simultaneously, as they complement one another. Step three builds from your insights developed in the first two steps. The other seven steps deal with developing, implementing, evaluating, testing, and refining restoration plans, based on how the ecosystem responds. Many of these steps can be pursued simultaneously, but we will caution you where needed to avoid getting the cart before the horse. Taking your time, working step by step through the process, achieves the greatest understanding and benefits to your restoration project.

### *Step One: Inventory and Mapping*

Begin by identifying different communities or ecosystems that occur on and around the property, the insults or *stressors* that are affecting them, and at least the dominant species present. Are there evident insults occurring to the ecosystem, and what consequences are they having? For larger acreages, these questions must be applied to each ecosystem or cover type, which we will call *management units*. Answers are revealed largely through personal interaction with the land; the ecological processes on it; and its residents, the biota.

Getting familiar with the land requires that you get out of your pickup and pull on your boots. It is impossible to see what must be seen initially even from the seat of a tractor. Only by walking, even getting on hands and knees to explore, can one assess indicators of land health and gain the needed perspectives. Find comfortable or interesting places to sit, watch, and listen. Especially go to areas that draw your attention. Often, these are the places with the most diversity and activity. It is no coincidence that more diverse places have the greatest appeal. Some argue that humans are genetically hardwired to forage for food and seek solace and beauty in nature. Such tendencies lead us to special places. Other places may also attract your attention: a large, fallen tree, a game trail, old fence lines, building foundations, or stone walls. Pay attention to streams and drainageways. How does water move across the landscape? Follow ridgelines, rock outcrops, and edges where different plant communities converge. These are spines that connect landscape features. If the land is all farmed, even the slightest change in relief or structure may draw your attention. In recently tilled fields, note changes in soil color and distributions of soil types. As you become more familiar with the land, features that initially attracted you will begin to connect to other features until, in time, you come to understand the landscape, whether it is largely desert, rangeland, or wheat fields, or a vacant lot in an urban setting. Get in the habit of carrying a notebook and perhaps a camera. Record what you see and the questions that occur to you. This includes recognition of different community types or land uses. These become management units, although as we get further into mapping and restoration planning, some may be combined.

**A good base map** is essential. It will be used for all planning. It should be proportionately scaled, and show relevant features or reference points on the land. For example, property boundaries, houses, farm roads, streams, internal and external fence lines, wetlands, and forests should be marked. Ideally, the base map will include the property boundaries superimposed over a recent aerial photograph (see fig. 2.1 as an example). All states now have digitized soil maps, and most have digital property plats that can be electronically superimposed over the most recent aerial photos. The local Natural Resources Conservation Services (NRCS) office may be able to provide this base map for you. In addition to the most recent aerial photograph, some agencies may have historic aerial photos that also will be useful (discussed later).

A topographic map also is valuable in the creation of a base map. Get one at the finest scale (usually 7.5-minute or 4-minute series appropriate to the size of area you are working). These can be obtained online from the United States Geological Survey (USGS). Both digital soil and topographic maps can be downloaded, with little computer expertise, from various Web sites or purchased from federal and state government sources. In most cases, you can obtain every map and aerial photograph needed from the Internet without charge. It is not very difficult to create a base map from online sources, but if need be, hire someone to do it. It will be money well spent. A consultant can use a geographic information system (GIS) or other specialty software to overlay these and other types of data (e.g., soils, hydrology, floodplain, surface geology, etc.) to create high-quality digital base maps. A benefit of having digital files is the opportunity to produce as many copies as you need at minimal cost and to easily change scale according to your needs. It also facilitates record keeping and easily allows addition of more data layers over time. You can certainly do this with hand-drawn maps, but they are time consuming.

The minimum elements of a good base map are topography that has your property boundary at the correct scale, recent aerial photographs, and soil types. Both the topographic and the soils maps will reveal surface drainage patterns. Add a north arrow and a scale bar. The scale of the base map should be large enough to allow you to accurately map small features, such as a clump of invasive thistles or the general location of a rare plant, or the nesting location of a bobolink, preferably no coarser than one inch equaling two hundred feet. However, if the property is greater than a hundred acres, a coarser scale may be required. Alternatively, break the property into units, and use a larger-scale map for each section. Print at least two copies of the base map, as those used in the field will become soiled. The base map will be used for mapping a variety of different features.

**Map ecological units (cover types).** After preparation of the base map, carry your notebook, a camera, and perhaps an aerial photograph, a topographic map, or other base map back to the field. On one of the base maps, begin sketching the cover types present on the land, keeping in mind anything you have learned about the historic landscape. A cover type is characterized by the dominant vegetation, and this should be resolved to the finest resolution you can map, usually, what you can see from a high-quality aerial photograph. Begin with the largest, most obvious, or most dominant features, such as woodland areas, pasture, streams, wetlands, buildings, cultivated fields, fence lines, and other features you find interesting. Be sure to examine any topographic oddities closely. They may be ecologically important because of species or unique communities that occur there.

Once you have a general map, you can begin to recognize further differentiation within a cover type. For example, you may be able to see different types of woodland communities within the forested area. Divide the woodlands into cover types based on dominant canopy species, or size and age of trees, or areas with infestation of non-native or invasive species such as chokecherry. Often different cover types or problem areas will coincide with different slopes and aspects. The aim is to generate an understanding of the structure and processes of each recognizable ecosystem or community, and its condition. This has to begin by identifying ecosystems in the field and mapping their location relative to topography, hydrology, and soils.

Be systematic in this inventory. Walk the property from one side to the other, or follow along the course of a stream from a lower to higher land, then through the uplands, reconnoitering the nonstream areas. Even on small properties, this process usually takes many days, and if you remain connected to the land, you will continue the process through the years. The focus of your attention will gradually shift from assessment to monitoring (discussed later) as the ecosystem regains healthy attributes. You may also wish to make lists of bird species seen or heard during these mapping exercises. Diversity of birds will reflect the diversity of the ecosystem in which they occur.

In wetlands, as in uplands, communities can be identified by dominant species: reed grass, reed canary grass, cattails, bulrushes, sedges, willows, alder, and plant communities, with more equal representation of many species. Map channels, open water areas, muskrat and beaver lodges, wild rice beds, or other features that catch your eye and interest.

Each geographic setting has its array of ecosystem types, and different suites of weedy plants that replace historic natural vegetation following different disturbances. In a forest, for example, you will want to note tree and shrub species, perhaps the distribution of tree sizes, abundance of spring flowers, snags and coarse woody debris, evidence of browsing, reproduction of trees, and presence of invasive species. In the southeastern United States, you might note where kudzu is smothering young tulip-poplar that colonized a former farm field, as evidenced by a plow layer in the topsoil. In western pastures you may find introduced crested wheatgrass and cheatgrass on what appears to be overgrazed prairie. Erosion will look similar anywhere in the world, but the plants that colonize such locations can range from persistent and aggressive weeds to rare native plants, such as prairie fame-flower, adapted to disturbances around badger diggings or bison wallows.

Even if many of the plants or birds are unknown, mapping can be done with descriptive notations rather than actual cover types. However, at some point, it will be necessary to become familiar with at least the dominant species, and sooner is better.

Among other data, collect plants that are unknown. Take good specimens, preferably with flowers or fruits, and press them in old books or magazines, or between sheets of newsprint weighted by magazines. Note dates and locations where they were collected. A botanist at the nearest institution of higher learning should be able to identify the plants, or direct you to someone who can help. Ask at the local extension office or library for help in finding books and other local information. You may find it especially important and useful to walk fields with a locally knowledgeable botanist. As you explore, keep a careful watch for rare habitats, such as overgrown prairie remnants or a spring hidden by invasive shrubs and reed canary grass. Neighbors, especially those who have lived in the area for a generation or more, can sometimes direct you to these gems in the rough.

**Describe cover types.** In addition to mapping, write descriptions of how each cover type looks, photograph it, list the prominent species, and list or describe the indicators related to ecosystem health. Are there invasive species? Is it diverse? Is it connected to similar ecosystems? Is there evidence of soil movement or erosion? If your property is a dozen acres or so, this can be done in a few hours. If larger, it may take many days or months to complete. One of the benefits of describing what you see is that this will force you to observe more carefully. The more you see and think about each community, the more insight you will have about how to restore it. Once the general mapping is completed and described, focus on problem spots and dysfunctional processes.

**Photo-document cover types.** A few photographs taken from the same location over time can, for example, provide a measure of stream-bank stabilization. A photograph of plants often will not be useful unless a list of plants, descriptions, or even measures of plant abundance (e.g., counting the number of ragweeds in a defined area) is included. On the other hand, a series of photographs of a patch of garlic mustard over several years, from the same perspective, can show progress (or lack thereof) toward control of this invasive species. Be sure to date each entry and note the location as specifically as possible. After you begin restoration, you will want to continue keeping good records on what was done and when. A record of what you have observed, or treatments made, and when, provide a good basis for understanding how different ecosystems or components respond over time. This is how the land advises us on proper treatment, and informs us as to whether we are moving restoration in the direction nature wishes to go.

Various types of forms are useful for recording data and information, or you can develop your own. We usually sequentially number the zones that are mapped based on the order in which they were investigated in the field and use this same numbering for the data forms. Fill in the date and other information on the data form. The more complete the information, the less confusing and more useful this process will be in the future.

## *Step Two: Investigate the History of the Landscape*

This step should be done simultaneously with the previous step. The more that is known about the history of the ecosystem and its relationships to people, the easier it will be to correctly diagnose the problems and prescribe treatment.[3] Using historic aerial photos, oral history gathered from older people familiar with the land, original land-survey notes, or other historic documents that may be available, map original community types on as much of the land as possible. What did the presettlement landscape look like and how was it changed from the way it was when only the Native Americans were involved with it? This provides the best indication of how nature developed the landscape before it was ditched, tilled, mowed, sprayed, pastured, or seeded to nonnative vegetation. Record the information you acquire in your notebook, along with any questions and hypotheses you have concerning each ecosystem. This will become increasingly important as you get further into the restoration process, especially if you are working with a landscape that has several management units or ecosystem types.

**Historic aerial photographs** are one of the best ways to obtain insight on the changes that have occurred in and around the property. Ideally, you will want the oldest aerial photographs available, pre-1940s for most areas in the United States. Most states are flown about every ten years, more frequently in many areas, so your current map should be no older than a decade. Using the oldest photo, trace a simple map of land-use types or vegetation cover. Note areas that were forest, cultivated, ditched, pastured, wetland, and so forth. Maps can be made with tracing paper laid over the aerial photographs, or digitally on a computer. A clear pattern of the changes that have occurred on the land, at least since the date of the earliest aerial photograph, begins to emerge when comparing these historic images with current conditions or the most recent aerial photo. Aerial photo interpretation requires some experience, so inquire at your local NRCS office for some help with interpreting the photos. Distinguishing between forest cover types, or an abandoned field and a hay field can be pretty subtle.

Look beyond the property boundaries, especially along drainageways. Even rough mapping beyond the boundary can reveal much about the trends and tendencies for land use. Neighbors tend to follow neighbors in the changes wrought on the land. For example, you may see ditching or channelizing of a meandering stream or drainageway that was straightened in 1956 on a neighboring farm, but not on your land. Then in the 1968 photo, you may see the channel extended through your property and to other neighbors up or down the drainageway. Watch for woodlands that were logged and converted to agriculture. Logging roads remain obvious for several decades if photos are taken in leaf-off seasons. Center-pivot irrigation units leave telltale circular patterns. When were forests cleared or fields first irrigated on or around your property? There are many changes you can identify, including some unique patterns that may be baffling at first and may require conversations with older people in the neighborhood to understand what happened.

Some of the discoveries may be exciting. We have worked on properties where Native American wickiup and long-house footprints were still visible from the rocks laid around the perimeter over two hundred years previously. For archeological features, look along slightly elevated rises on the downwind side of rivers, lakes, and wetlands, where access to water would have been present but the location was protected from the wildfires that were pushed by prevailing winds. The patterns of change in rivers can be a study alone. Rivers can change course quickly, especially in agricultural landscapes. All meanders, oxbows, and sand and gravel deposits across a floodplain were previously occupied and created by the river. Your tracing should note these features so that you can quickly focus on those that are different on the land now.

**Identify reference natural areas.** If available, obtain a directory of state natural or scientific areas from your state natural resource agency. Determine the areas nearest the geographic location of the land being restored, and visit them. Cross-reference the soils and topography, and see if vegetation (perhaps even detailed species lists) has been described. Matching plant communities present in nearby natural areas with the soils, hydrologic, and geologic settings of a property to be restored can provide useful information on species and community types that may once have occurred there.[4] These should correlate with the original survey data described earlier. For example, if the survey description indicated an oak savanna, then the natural area at that location should be an oak savanna. If it is not, then either there were errors in identifying the location, or more likely, changes in disturbance history have resulted in a change in the vegetation, thereby reducing the value of the area for your purposes.

After Steve saw the possibilities in local natural areas, he found fragments of similar prairies, savannas, and wetlands still remaining on Stone Prairie Farm. This information was directly applicable to development of working understandings or models of what potentially could be restored on his farm and what species should be included.

**Original land surveys.** Thomas Jefferson established the General Land Office (GLO), and initiated surveying all U.S. territory. In addition to providing a basis for establishing land ownership, the process left an early, if sketchy, record of ecological systems on the land. The survey laid out a systematic, hierarchical land division. Counties were aggregations of areas called townships, each thirty-six square miles in size, which in turn comprised sections, each a mile square. Each section contained 640 acres. Sections could be easily divided into quarter-sections, each 160 acres, or quarter quarter-sections, each 40 acres. Surveyors dragged steel tapes, called chains. Eighty chains equaled a mile. Individual links on the chain were just less than 8 inches (7.92 inches) and 100 links equaled one chain.

Surveyors pulled their chains along compass lines, noting quarter-section points halfway along each section, and establishing section corners, referencing location to nearby trees, which they identified by species and diameter. They also recorded features encountered along each pull, but primarily focused on vegetation transitions of economic value. For example, they would record where dry soil vegetation along rocky ridges gave way to vegetation indicating loamy soils better suited for farming. They also noted where scrubby timber transitioned to trees that would produce better lumber. Water resources were also commonly noted, undoubtedly to help identify potential for livestock production. Other useful observations included poorly drained areas, lakes, ponds, and streams. Their notes sometimes included reference to game observed, or where fires had recently burned.

In Wisconsin, the earliest surveys date from 1837 to the 1850s. They provide a thumbnail sketch of approximate locations of major vegetation types and where wetlands or stream channels were located (and their widths, if the surveyor had to cross them) at the time that the original surveys were conducted. Maps of the original vegetation of Wisconsin have been prepared from original survey data. Similar resources are available in many other states. While coarse, these data can give you a good idea of the kind of presettlement vegetation on your property. In addition to vegetation types, locations of streams, lakes, wetlands, farm fields, residences, river fords, and other features encountered during initial surveys can be learned.

Continue to record questions stimulated by your observations: How long was this area pastured? What was here before pasture? When was it cleared? If this field is tiled, when were tiles installed? The farther back you can evaluate the history, the more insight you will gain about the natural processes that gave rise to each community. (We later will go through a more formal process of assessment, interpretation, and the restoration design where these details are especially useful.)

Your observations will lead to other questions, depending on your experience. You may wonder how to stop some erosion. What should be done about a ditch that was cut deep into muck soil that could have been developed only under wetland vegetation over thousands of years? Why are there such deep gullies on a hillside when it appears there is little current erosion? Why are there boxelder trees under the oaks in that forest? Make notes of all questions. They are important. In time, answers for most will be revealed, or professional assistance can help interpret the clues.

It will be necessary to walk the land several times during different seasons to note activities or species not evident at other times. This, however, does not delay you from beginning the next step.

## *Step Three: Interpretation of Landscape Changes*

Having walked the land a second and third time, you will have many communities outlined on maps and numerous notes and questions in your notebook. You may have discussed observations and questions with local residents or naturalists. Now it is time to pull that information into a usable format so that you can begin developing restoration plans. These plans will remain flexible and likely will require frequent revision. As projects are implemented, and as the land responds, you must also respond. Ecosystems are dynamic and responsive, and the way they respond to restoration treatments is one of the most important additional sources of information.

Your curiosity eventually will lead to insight into how to improve land health. Start by thinking about the disturbed and damaged areas. Some may be so damaged that it is not clear where to start, while others are obvious and can be tackled with limited resources, like removing junk from a spring, or fencing cattle out of wetlands.

**Mapping stressors.** It is important to recognize what has changed, how, and why. Most of the changes will be associated with physical, biological, and chemical factors that negatively influence the condition and ecological health of your land. These are the stressors. Areas being impacted by various stressors, regardless of source, should be described and mapped. Even speculative information about what may be impacting each ecosystem will be useful.

Some stressors originate off-site. Common examples, easily seen in agricultural areas, include herbicide drift (e.g., herbicide carried onto your land by wind or water), erosion and sedimentation, and storm water runoff. Mapping should indicate the direction and off-property locations where erosion and drainage are believed to originate. Place approximate boundaries around the sediment deposits. Often the boundaries will coincide closely with the growth of one or more invasive plant species. Fertilizers carried in runoff or sediments from neighboring farms can stimulate growth of weedy plants to the detriment of more native vegetation. Nonnative or invasive species may be carried from adjoining land, spread by wind, water, game, or livestock. Acid precipitation is an off-site stressor impacting forests throughout much of the East, more so where soils are thin and coarse. Internal stressors may be erosion, where topsoil has been lost; altered disturbance regimes; damage from off-road vehicles; invasive species; altered hydrology; or overgrazing. Less obvious examples include locations where ditches or tiles release water, or where high-volume groundwater irrigation pumps may reduce water in wetlands. Former wetlands will often be indicated by muck soils that were formed under saturated soil conditions. Many are now hidden by agricultural crops.

Also note any direct intrusions, where activities on adjacent land crossed property boundaries. Examples are trees cut on the property, off-road vehicle trespass, or invasive weed populations that are spreading onto the property. Map access roads (trails, logging roads, etc.) that may be corridors for invasive plant species or unwanted trespass. Note any places where neighboring livestock find their way onto the land. Streams present another suite of concerns. Nonnative species often migrate along stream corridors or in the stream itself, so awareness of problems both upstream and downstream can be important. Water-quality problems can originate on the property as well as upstream. Flood regimes are controlled by changes in upstream hydrology as well as by climatic shifts.

**Identify and map ecosystem dysfunction.** Once the resources are mapped, and the off-site and on-site stressors are mapped and described, you can begin to define and map the ecological health within the landscape. This can first be done at a general classification level: healthy—no particular problems or concerns; moderate problems—not healthy, but with basic ecological processes functioning; severe stress—loss of ecosystem structure and functions. You also sometimes identify areas of uncertainty, where testing and experimentation may be needed to elucidate problems or ecosystem dysfunction. This exercise should be done at as high a resolution as possible. A separate map for each management unit is often better.

One approach is to overlay tracing paper on the base map and color areas with different categories of stress and associated dysfunction. For example, you might color areas yellow that have little or minimal topsoil because of erosion. If erosion is a widespread problem, you might separate moderately eroded areas (yellow) from severely eroded areas (red). Areas of soil deposition, usually with nonnative vegetation, might be colored brown. Dewatered wetlands might be light green if moderate, and dark green if severe. Patches of nonnatives could be light blue if minor, and dark blue or purple if severe. The aim is to partition each management unit into zones that represent ecosystem health, with as fine a resolution as possible. Because few if any areas will be perfectly healthy, it may be helpful to think in terms of ecological problems. Except for very pristine natural areas, every area will have some problems, ranging from so severe that few natural functions or structure are present, to a relatively healthy ecosystem. The latter might be the case in high-quality natural forests or wetlands; an example of the former might be a field that is being row cropped with use of cultivation, fertilizers and pesticides, often surrounded by ditches or underlain by tile, or irrigated.

**Map changes and the magnitude of change.** Now, you need to summarize your maps and notes to create a new map that identifies changes that have occurred since presettlement in each zone you previously mapped. You may want to do overlays, with one map indicating vegetative changes, another drainage changes, and so forth. In the most general way, indicate extreme, moderate, or low categories of change. Low change for soils might be where they have not been plowed, but the area has been heavily grazed. Moderate change is a woodland where timber cutting and grazing have changed the species composition or allowed invasives to become well established. Extreme change is usually associated with intensive agriculture or development, such as draining a wetland or cultivating a dewatered field.

**Evaluate the possibility of restoration and potential restoration success.** Now it is time to begin making restoration decisions. The information you have collected and mapped will provide the basis for highest priority restoration as well as management strategies. Simultaneously, in this same process, you need to identify dysfunctional areas you are unable to address at this time. Some require your neighbors' participation. Some of the problems and stressors are just too big to even contemplate tackling initially. Mark these on the base map and in your notebook as not currently solvable or addressable. Make sure in the data form that areas that cannot be currently addressed are well defined with boundaries and the nature of the problems are clearly identified.

Using information gathered to support your observations, you can get some ideas of what the healthy ecosystems on the land were like before they were damaged. These models, together with what you learned about ecosystems before settlement, or that still exist in nearby natural areas, inform you of the potentials for restoration and serve to guide the restoration process. You nearly always want to move a system toward one that is self-maintaining, usually what nature initially created in that location.

Using the ecosystem health map, begin to identify areas that are most easily restored. Ideally, some areas may need no treatment at all. In others you may only need to restore some native species, remove cattle, backfill ditches, or disable tiles. Also, identify areas where restoration success, given current resources, is likely to be limited. These may be areas where experimentation and on-the-ground testing are needed to access the possibilities, or test alternative approaches. It is the areas that have the highest potential to achieve restoration success where restoration programs usually are implemented first. Consider all possibilities, then establish some priorities that are incorporated into annual planning in logical sequence.

In addition to off-site and on-site stressors, there also are practical constraints such as funding, labor, and your energy that may make it difficult or undesirable to undertake some projects, at least initially. Always try to map the best-case possibility, and then reserve for future reevaluation areas that cannot be restored because of stressors and other constraints. Always be very clear, however, why the areas cannot be restored.

**When have you gathered enough information?** Getting to know a particular landscape is a lifetime process. The important question, therefore, is, "What details and what knowledge are indispensable to getting started with restoration?" The simple answer is, "Enough to make good decisions." Without at least some understanding of how the land has changed and why, you cannot be sure that your management decisions and restoration will begin to restore ecosystem health. While some people are tempted to hold back and delay the restoration process, others may be tempted to take action too quickly. We have been torn between both—indecision leading to inactivity, and assumed understanding leading to premature action. We have learned from both, and this has helped us become better at the process of discovery, inquiry, recognizing solutions, and implementing some tasks with a sequenced, methodical approach.

A simple test can guide decisions: Will the proposed restoration activity likely result in greater native diversity? If so, it probably is a good decision, or at worst, one that will not lead to greater harm, although there are exceptions. The exceptions are generally where the ecosystem is very healthy with no serious problems. When in doubt, hesitate and dig deeper for understanding.

It often is necessary to apply the same test to advice from others, even professionals. Steve consulted the local NRCS office for suggestions on stabilizing the stream bank at Stone Prairie Farm. Their solution, offered without a field visit, involved reshaping the existing stream channel, then dumping twenty to thirty loads of coarse stone from a local quarry to stabilize banks. They then proposed planting exposed soil with reed canary grass! Luckily, Steve knew that reed canary grass was aggressive and probably would prevent establishment of native diversity along the stream. He developed an alternative plan that was cheaper and more ecologically sound. A local contractor was hired to regrade the banks that were then seeded with native wetland species and short-lived annual ryegrass as a cover crop. Within a year, diverse native vegetation grew over the stabilized banks, without rock or reed canary grass. Now, more than twenty years later, banks remain stable and are lush with native species. Had Steve followed the NRCS recommendation, he would have qualified for federal matching funds but would have had undesired results.

Especially with ecological restoration, professionals often have inadequate training or experience and too narrow a focus. Forestry consultants, for example, may be more focused on producing timber products than on ecosystem health. Fisheries consultants may be more concerned about water conditions from the standpoint of healthy sport-fish populations, not necessarily healthy stream, lake, or pond life. Therefore, ask questions, seek opinions, but be cautious. A red flag should always go up if the suggested solution seems to focus on favoring one or a few species rather than restoration of native diversity. Always work toward restoration of complete ecosystems.

## *Step Four: Develop Realistic Goals and Objectives*

For each management unit, based on an assessment of the potential for restoration, develop management goals and objectives, with timelines, and list the resources required for the project. Perhaps some units will be kept in present condition for the time being, or maybe resources and time are lacking to address the more severely degraded units. Remember that the healthier the ecosystem, the easier it is to restore. Establish priorities for restoration based on time and resources, and the potential for restoration. These thoughts and ideas should be noted in your notebook.

Priorities throughout the restoration and land-management planning process, and over the years of implementing the program, will change depending on how the land responds to management, and how you respond to the demands of restoration. Nevertheless, it is important to develop goals early in the restoration process. Goals help you stay focused and provide targets. Often, especially at the beginning, you may be overly optimistic or ambitious. If so, or if the land does not respond as anticipated, underlying assumptions must be reconsidered, and goals adjusted. Likewise, restoration methods and strategies may need to be refined.

When applied to a restoration plan, goals can be both philosophical and pragmatic. Pragmatic goals should be measured against successes and failures. Be aware that goals can be incompatible. For example, you may want to manage deciduous forest for presettlement conditions and exclude hunting. Dense populations of white-tailed deer, commonly five to ten times more abundant than during historic conditions, preclude the compatibility of these goals. Good examples of goals for restoration might include such things as reducing nonnative species, restoring surface and subsurface hydrology, and creating educational opportunities to increase awareness.

## *Step Five: Prepare a Plan*

For each unit, set priorities so that the most critical tasks are done first. For example, Steve recognized the need to control erosion and nutrient input to the stream on his farm before much could be done to improve the stream ecosystem. There also may be reasons certain tasks are done earlier, having to do with ecosystem processes, available funds or time, or simply for some quick successes before laboring toward goals that will come more slowly. Steve started prairie plots to provide seed sources for when he scaled up the restoration of prairie on his farm, for example.

### *Step Six: Develop and Initiate a Monitoring Program*

A monitoring program (the focus of chapter 5) is essential. A good monitoring program will allow you to evaluate progress toward your goals.

### *Step Seven: Implement the Plan*

Professionals often prepare annual plans that indicate what and when different tasks will be scheduled in the coming year. This can be helpful for even small, backyard restoration projects. Once a plan of action is completed, begin to line up necessary resources to accomplish tasks scheduled for that season.

### *Step Eight: Document Changes and Maintain Records*

We recommend preparing annual reports that also explain the project, actions taken, and summary of results. This step is usually essential for professionals, but can be very helpful for any restoration project. Often done at the end of a season, it provides an opportunity to evaluate what was accomplished, progress made toward goals, reexamination of hypotheses and approaches, and completion of one cycle of adaptive management. From this perspective, begin preparing next year's annual plan. The report also provides a permanent record of the work completed. It may also be a way to file documentary photos. For simple projects on your own land, the annual report can be as simple as keeping a journal on observations and work activities.

### *Step Nine: Periodically Reevaluate the Program*

Especially as new information and ideas occur, review the restoration program. An ideal time to do this is in conjunction with the annual report. After several years, you will have accumulated a wealth of information about how various units are responding to restoration treatments. In addition to annual review, the overall program should be periodically evaluated.

## *Step Ten: Communicate and Educate*

People who are nearly always interested and potentially affected by the restoration include neighbors, extended family, and volunteers. Connecting with nature is rewarding, exciting, and fulfilling. Share the experience. The rewards of doing so will be cooperation, assistance, and the education of others. We often are asked by local school groups if we will lead a field trip to a restoration project, and we always try to oblige. When possible, we arrange to have the group get some hands-on experience, such as checking bluebird boxes for nesting success, pulling nonnative weeds, collecting seeds, or doing some simple stream monitoring. A local bird club in central Wisconsin purchased a natural area in its town and began restoration. They improved trails and put up nesting boxes to encourage more species of birds. Teachers at a local grade school were looking for a nearby field trip location and began visiting the area. Members of the bird club asked the school to help restore the property, and the school adopted it as a project, providing hundreds of extra hands to help and a great learning opportunity for the children. A large display with photographs in the school lobby is regularly updated to keep teachers and students apprised of the progress.

# Conclusions

The restoration process begins with personal observations, then applies information from all available sources to develop an understanding of ecological health on your land. Through your understanding, you identify and map the restoration needs and opportunities, and set priorities. Products of the planning include the following:

1. Maps of cover types, ecosystems, and/or current uses.
2. Map of constraints present on the land or imposed through neighboring land uses. We have referred to these as stressors.
3. A map of historic cover types.
4. A map showing ecosystem quality, or health zones, or zones of dysfunction.
5. A map of restoration potential.
6. Draft of restoration goals and objectives for each management unit.

Completing these products will provide you a good understanding of your land, its history, its stressors, and what you hope to accomplish during restoration. In the next chapter, we expand and explain these steps further with specific examples from the restoration work at Stone Prairie Farm.

## Chapter 3

# Developing a Restoration Plan

*Failing to plan is planning to fail.*

Alan Lakein

The more you invest in restoration planning, the more valuable a plan becomes. Planning for restoration at Stone Prairie Farm required two years of thorough investigation and some experimentation, not unusual for a sound restoration project. Planning always begins with analysis of existing conditions and an investigation of how the land has changed over time. It requires that you develop a personal relationship with the land. Hesitantly, we suggest that this process is a love affair between you and the land. As you come to know the land, you become increasingly attached and more concerned for its welfare. Greater humility follows, and that opens you to new awareness and learning opportunities. The restoration plan can be no more static than your feelings for your land; it must continue to evolve and mature.

Ecological restoration, properly done, is a systematic process whereby each step increases understanding of the ecosystem. Because you never fully know, nor can you anticipate, how complex ecosystems will respond to any particular treatment, the plan must remain amendable as information is gained. In chapter 2, we covered the basic framework of the restoration process. In this chapter, we use the restoration planning process for Stone Prairie Farm to illustrate the first six steps in the ten-step process. (In later chapters, we will illustrate the application of the process in other geographic settings and ecosystems.)

A good restoration plan should accomplish two objectives: guide investments of labor and dollars to most efficiently implement the plan; and provide accurate records of what was done so that you can learn and adaptively refine or change your approach when the land responds differently from what was expected, or when resources require modifications. Plans lose value if basic steps and products of the foundation procedures are neglected.

As with a building, four good cornerstones are required to create a strong, durable foundation. In a restoration plan, the cornerstones are the accuracy and depth of the interpretation of existing conditions, historic conditions, causes of changes that have occurred, and goals that reflect the inherent capacity of the ecosystem. Keep in mind that in large projects, these cornerstones will likely differ for each subunit of the area you are restoring, but they should always be thorough. Also, when developing goals, some frank discussions or soul-searching is needed to determine how much personal time and money you or others are prepared to invest, and what results you need to stay invested and motivated to continue. We will expand on this in the next chapter.

At Stone Prairie Farm, the original planning and mapping were all completed with pencils and tracing paper. Steve used a photocopy machine to enlarge or reduce reference maps to a common scale. When initial planning was done in the early 1980s, digital-mapping technologies were not available. Steve chose the scale of one of the available maps that seemed to give the greatest flexibility and utility and superimposed other mapped information at the same scale. Some figures in this chapter were originally developed using that low-technology process, but they have been updated with geographic information system (GIS) technology. Either approach works. Although there is more precision and flexibility when using computer-assisted drafting (CAD) and GIS, the values of the maps are dependent on the quality of the content that is mapped, not the method. For this reason, we neither encourage nor discourage any particular technology but focus on the process and thinking in the restoration process.

## The Restoration Process

Refer back to chapter 2 if necessary to refresh your understanding of the steps in the planning process. In this chapter, we will illustrate the activities involved, primarily with reference to Steve and Susan's planning and restoration of Stone Prairie Farm. Keep in mind that especially steps one, two, and three can be, and often are, pursued simultaneously.

### *Step One: Inventory and Mapping*

Start with the base map prepared as described in chapter 2. Each task involves compiling specified information on a new sheet of tracing paper that you lay over the properly scaled base map. During this process, choose the color or symbol scheme you will be using for each sequential layer of information. You can include more or fewer layers of information on the same piece of tracing paper, as long as the clarity of information on the map is not compromised.

**Existing conditions.** Prepare a scaled map of the existing cover types on the land. Additionally, map directly abutting lands that may influence the property. For this step, use whatever nomenclature scheme for naming locations and cover types you wish, but be sure to document and describe each in your notebook so that you can be consistent throughout.

At Stone Prairie Farm, Steve defined ten cover types, numbered sequentially from the most abundant to least abundant (figs. 3.1a and 3.1b). Type 1 areas were cultivated farm fields growing corn.

Type 2 was forage/hay fields. Type 5 was pastured riparian wetlands, nearly all of which were badly eroded and highly degraded with box elder and white mulberry thickets along lower areas of the stream course. Upper stream locations were dominated by stinging nettles and other invasive weedy plants. A dry south-facing gravelly ridge that contained a gravel quarry was mapped as type 6. Steve thought it was probably a dry prairie remnant because of the presence of some native plants such as little bluestem and wild rose. Type 8 was the developed land around the farmstead itself, with ornamental landscaping, some fruit trees, lawns, gardens, and orchards. When the mapping was extended to adjacent land, two additional cover types not present on Stone Prairie Farm were added.

At Stone Prairie Farm, populations of invasive plant species, especially those species that are aggressive, and persistent perennials, were also mapped. Abbreviations were used to indicate where invasive species occurred. Nonnative species were more common than native species in all cover types except the hill prairie, so only the most severe infestations were mapped.

**Hydrology.** Surface hydrology is most easily mapped over a topographic base map, although it can be done with a good aerial photograph. Map all drainageways, including areas where precipitation and snowmelt move as sheet flow. At Stone Prairie Farm, Steve realized that several different hydrologic features were present and needed to be distinguished on the map. He recognized a perennial stream, intermittent drainageways, springs, and tile outlets (fig. 3.2). One also might choose to distinguish between different size streams or gullies, ephemeral and perennial springs, or other differences that seem important.

Overland sheet-flow locations were easy to map by walking through the cornfields. Beneath the corn, evidence of sheet erosion of topsoil was conspicuous and revealed the overland routes that water followed as it ran off the land. Sheet erosion was shown on the map with arrows and cross hatching.

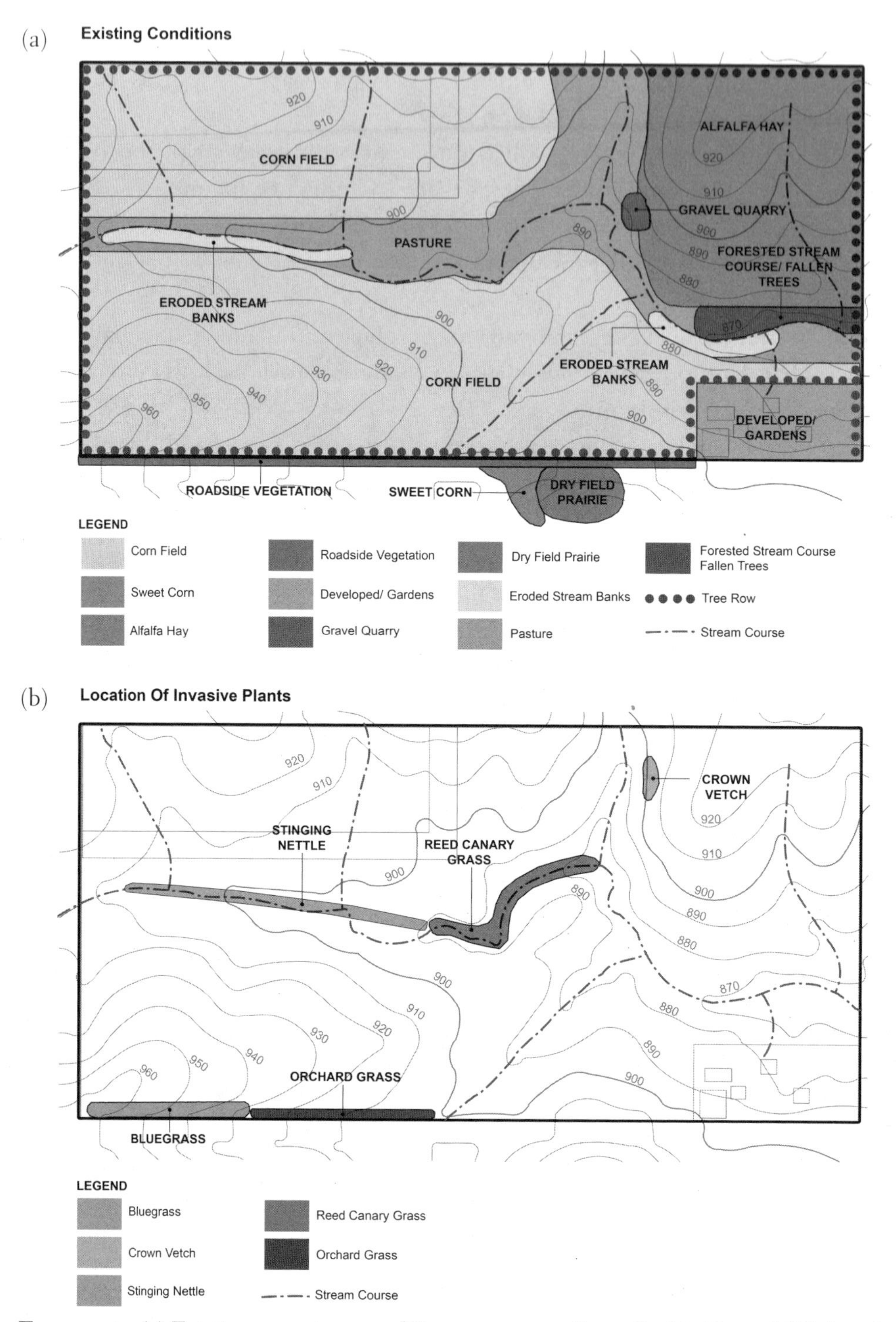


FIGURE 3.1 (a) Existing ecosystem conditions present on Stone Prairie Farm, 1981, based on mapping by Susan M. Lehnhardt and Steve I. Apfelbaum. (b) Location of invasive plant populations during the survey of existing conditions, 1981.

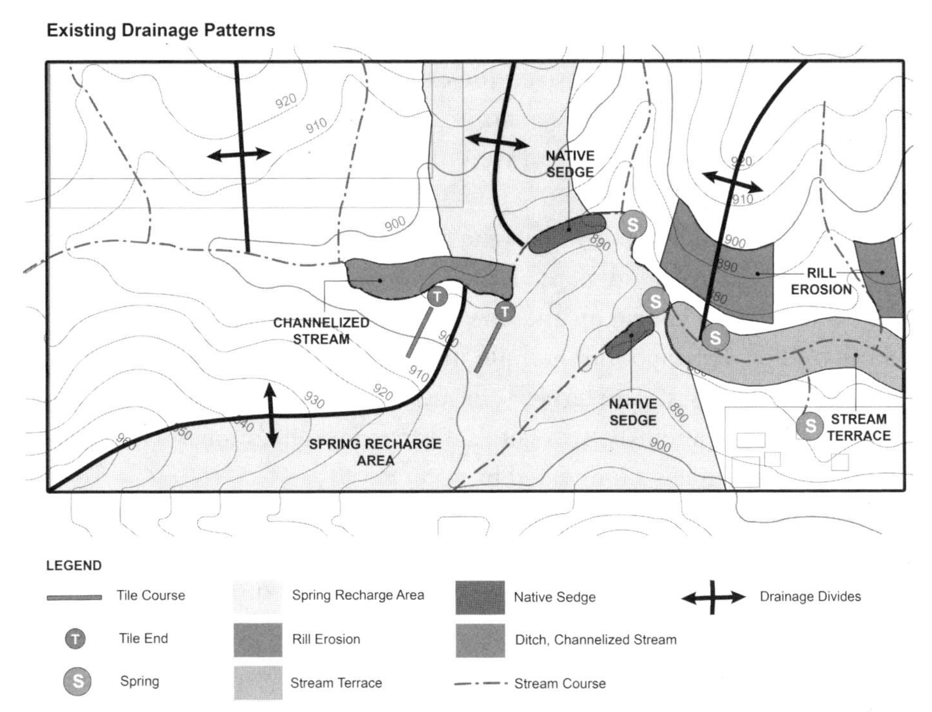


FIGURE 3.2 Existing drainage features and hydrology present on Stone Prairie Farm, 1981.

Upland areas that were thought to be important groundwater recharge zones, where precipitation seemed to infiltrate, were also mapped. Years later, after the stream recovered, Steve and Susan found many more springs that were not conspicuous during the original surveying. Many may have become active as a result of increased infiltration following restoration. The springs initially mapped were only those with the highest discharge that were able to bubble up through the mud, algal mats, and turbid water where the cattle had damaged the stream. As others were noticed over the years, locations were added to the hydrology map.

Steve also mapped the surface-drainage divides. This is most easily done with a good topographic base map. He walked ridge lines on his and neighboring properties to confirm these boundaries, and located culverts beneath the roadways that carried water into Stone Prairie from neighboring properties.

In parts of the East and Midwest in particular, agricultural drainage tiles have been extensively used to dewater agricultural fields. Tile outlets are often concealed by vegetation or cattle disturbance along drainage ditches or streams. At Stone Prairie, presence of tiles was indicated in some locations by broken clay chips along the margins of the stream. Steve and Susan mapped the outlet of these tiles. Sophisticated equipment is available for finding and mapping the locations of tiles, but

Steve learned the art of finding tiles by "witching" or "dowsing" them. While theories abound, we know of no good scientific explanation for how dowsing works. Tile lines were indicated on the same map as other hydrologic features. Tile outlets where lines emptied into the stream were distinguished from tile lines located by witching and digging.

When property does not encompass the head of watersheds, as with Stone Prairie Farm, it is important to extend maps of some hydrology features to neighboring properties. Map the stream course up and down the watershed for some distance. Other features that might be important include intermittent drainageways; ponds; ditches or channelized streams; stream terraces, formed by the deposition of waterborne sediments during floods; and oxbow wetlands. If streams on the property flood, it is useful to note the areas that are inundated by water.

**Soils.** Soils can be both the easiest and most difficult features to map. Because soils in the United States have been mapped by the USDA Soil Conservation Service (SCS), beginning in the 1930s and refined through the 1970s, generalized soil maps are readily available (fig. 3.3). SCS was reorganized into the Natural Resources Conservation Service (NRCS), which continues to assist landowners with soil and water conservation projects. Similar information is available in Canada through the Cana-

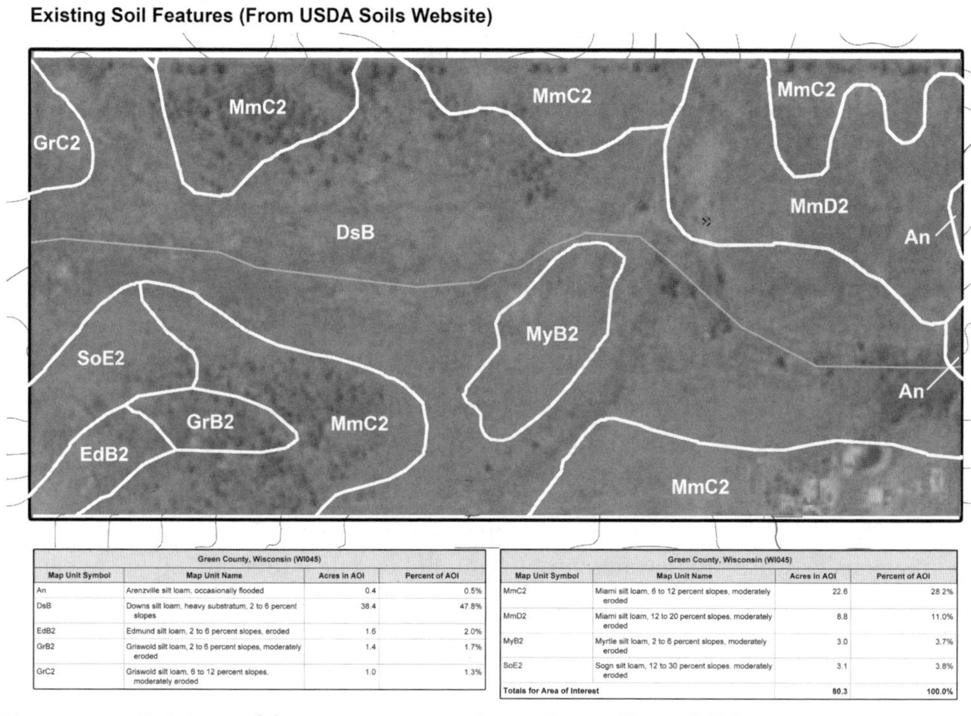


| Green County, Wisconsin (WI045) | | | |
|---|---|---|---|
| Map Unit Symbol | Map Unit Name | Acres in AOI | Percent of AOI |
| An | Arenzville silt loam, occasionally flooded | 0.4 | 0.5% |
| DsB | Downs silt loam, heavy substratum, 2 to 6 percent slopes | 38.4 | 47.8% |
| EdB2 | Edmund silt loam, 2 to 6 percent slopes, eroded | 1.6 | 2.0% |
| GrB2 | Griswold silt loam, 2 to 6 percent slopes, moderately eroded | 1.4 | 1.7% |
| GrC2 | Griswold silt loam, 6 to 12 percent slopes, moderately eroded | 1.0 | 1.3% |

| Green County, Wisconsin (WI045) | | | |
|---|---|---|---|
| Map Unit Symbol | Map Unit Name | Acres in AOI | Percent of AOI |
| MmC2 | Miami silt loam, 6 to 12 percent slopes, moderately eroded | 22.6 | 28.2% |
| MmD2 | Miami silt loam, 12 to 20 percent slopes, moderately eroded | 8.8 | 11.0% |
| MyB2 | Myrtle silt loam, 2 to 6 percent slopes, moderately eroded | 3.0 | 3.7% |
| SoE2 | Sogn silt loam, 12 to 30 percent slopes, moderately eroded | 3.1 | 3.8% |
| Totals for Area of Interest | | 80.3 | 100.0% |

FIGURE 3.3 Existing soil features present on Stone Prairie Farm, 1981.

dian Soil Information System (CanSIS). Soil information is available for Mexico through the North American Soil Properties (NOAM-SOIL) project, although the data are too coarse to be of much use at the individual project scale. In the United States, soil maps are digitized and available online for many states. Check with your county extension or NRCS office for where to order a hard copy of the soil maps and descriptions. At Stone Prairie, Steve and Susan soon discovered that even detailed soil maps were often inaccurate or lacked important details. Correcting and refining a soil map proved to be one of their greatest challenges.

SCS did most of their mapping from aerial photographs with limited ground-truthing or field checking. Although no field confirmation had been done on Stone Prairie Farm property, the SCS map captured the broad generalizations reasonably well. SCS maps are usually sufficient for agricultural use where tillage, fertilizers, and agronomic crops are less sensitive to minor variations in drainage and nutrient retention. In natural plant assemblages, the diversity of vegetation reflects even small differences in soil characteristics. Thus, to facilitate restoration of natural vegetation, you need to know more about variation in soils than is generally indicated on an SCS map. Steve and Susan enlarged the SCS map to the same scale as the base map and then transferred the boundaries of each soil type onto tracing paper. That was the simple part of this task.

With a soil probe purchased from a science-equipment supply company and a sharpened shovel, Steve and Susan started probing and digging. They read the descriptions from the SCS county soil survey and examined the soils, trying to understand the discrepancy between what they saw and what the SCS map indicated they should be seeing. Clay loam soils, clay being the predominant soil-particle type present, were found along ridges. In other areas they found silt loams, where the larger and less cohesive silt particle size dominated. The general soil map failed to recognize that often on the same ridge, and within several hundred feet, both would be present. Clay loams and silt loams are quite different in the rate that water can infiltrate, or how well they hold water and nutrients. Natural vegetation clearly reflects such differences. In a cornfield, these differences are much less apparent.

The uppermost layer, or topsoil, called the A *horizon*, in the clay loam was relatively deep where it had not been eroded from the land. In some instances, the surface was eighteen to twenty-four inches of dark, organic-rich soil. In contrast, in silt loams, the A horizon was lighter and consisted of only a few inches over the subsoil. It was often integrated with sandy loams that sometimes had no discernable topsoil at all. In time, Steve and Susan began to realize the importance of these observations. These soils indicated the transition between historic prairie ecosystems that developed in the clay loam soils and savanna and forested ecosystems that occupied the silty and sandy soils. Stone Prairie Farm had not been all prairie but had been a patchwork of prairie, savanna, and forest. These ecosystems had expanded and contracted, no doubt with climatic fluctuations and fire frequency and intensity. Thus, the soil map now provided much additional information on the community types that occurred, and even some insight into the dynamics of these ecosystems.

Fortunately, in many areas, this much detail in soil mapping will not be necessary. Southern Wisconsin lies in an ecotone, where tallgrass prairie, oak savannas, and eastern deciduous forests came together and intermingled. If your property occurs in an area where there is much variation in drainage or topography, you may need to spend considerable time refining the soils map, but in few areas will it require the attention to detail that Steve and Susan invested.

At Stone Prairie, Steve and Susan also found discrepancies in the SCS soil map along drainageways and in the lower ground. Using a soil probe and shovel, they could quickly determine where the black, largely organic mucks indicated saturated soils, or where wetlands occurred historically, before they were tiled and drained. Muck bordered perennial drainageways, where they also found springs. Some historic wetlands were even found along upland slopes, indicating where historic springs had once flowed.

The soil mapping provided other important insights to the history of the property. It revealed where the channelization process along a stretch of the stream deepened the channel through a historic wetland. The excavated ditch captured the stream and drained the wetland. The deep muck of the former wetland now supported mostly nonnative species, corn where tilled, and weeds in the pasture. When it was later witched, numerous tiles were found in the area.

By probing, Steve was able to find where the historic stream channel had been. Because some ditching predated the first aerial photos, he had to determine the historic channel location by the presence of heavily mottled clay deposits, mostly light and dark gray colors indicating reduced iron, caused by exclusion of oxygen. The location of the historic stream channel was located on the soil map and also on the hydrologic map.

Steve also learned through this process where former topsoil had eroded from the higher slopes. He was alerted to this in areas where the description indicated that there should be eighteen to twenty-four inches of A horizon, but there were now only a few remaining inches. Often, on the lower slope beneath, he could find buried profiles, where the eroded topsoil was deposited over the soil profile that had developed under the historic vegetation.

**Anthropogenic features.** Many features on landscapes were created by humans (fig. 3.4). Some are ecologically important, and others of historic interest only. At

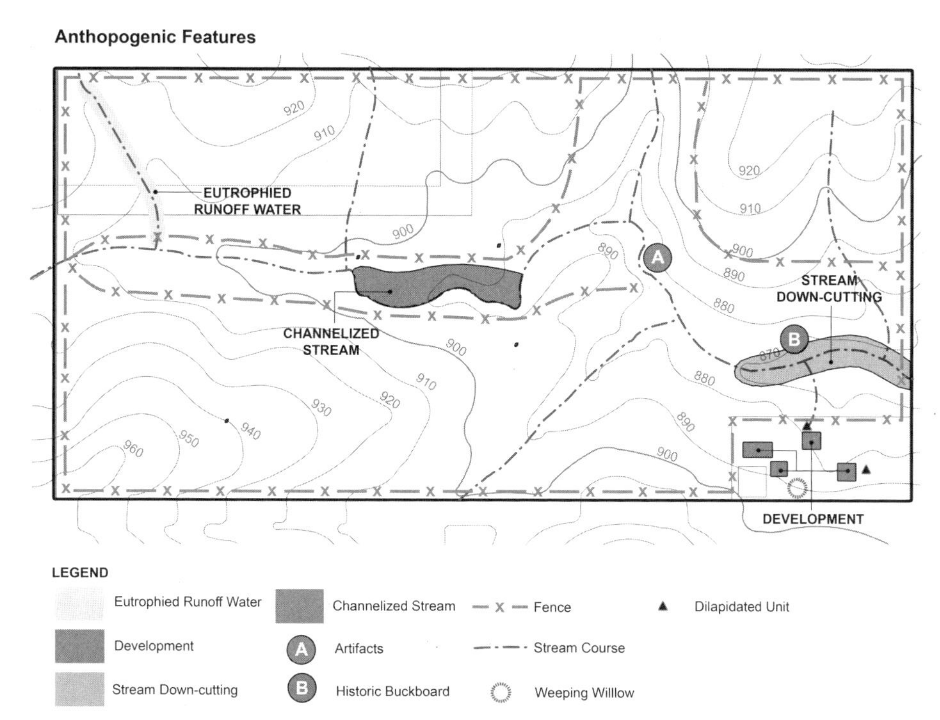


FIGURE 3.4 Existing anthropogenic features present on Stone Prairie Farm, 1981.

Stone Prairie, Steve mapped fence lines; trees that had been planted around the farm buildings; the channelized stream; water well; piles of rock; old cement foundations marking the locations of long-gone buildings; waterways constructed to intercept runoff from neighboring cattle operations; even a piece of an old wagon eroding from the stream bank; and exposed trash dumps with old farm implements, cans, and bottles.

Steve was cued into a rich history with the discovery of the hames that would have attached leather harnessing to necks of draft horses in the early farming days. One hame eroded from the stream bank, and a second, a match to the first, was plowed up in the farm field just above the stream as the land was being readied for planting the year he purchased the farm. In discussing the hame with the farmer, Steve learned that the man knew a lot about the history of the land.

The farmer had found arrowheads near the larger springs, an obvious testament that the Native Americans also valued fresh, cold water, and perhaps camped or at least hunted around the springs. Steve began looking around the springs and found many chips of chert, but it was Susan who found a prized piece: a stone knife, probably a hide scraper. It fit the hand perfectly, with a place for the thumb and index finger to apply guidance to the sharp blade. It helped connect Steve and Susan to the people who first lived on the land and appreciated its natural qualities. They mapped where artifacts were found.

While the abundance and kinds of anthropogenic features found will vary from one site to another, it is likely they exist in nearly every area where ecological restoration might be contemplated or required. On your property, you will need to decide how important these features are and to what extent you may wish to preserve them. Presettlement features created by Native Americans should be carefully mapped and preserved.

**Potential seedbank.** Where are weeds or native species likely to emerge from the soils? This information can prove quite useful. Steve and Susan collected samples from the upper six inches of soil in representative locations. They scattered the soil samples over flats of sand, and allowed plants to emerge and grow until they could be identified. They learned that disturbed muck in some of the lower areas along the stream contained mostly invasive plant seeds, especially stinging nettle and reed canary grass. They also discovered, however, that some native wetland species remained in those soils.

Only agricultural weeds were found in the clay and silt loam soils from the ridges that had been so badly eroded over many years of farming (fig. 3.5). Areas along the

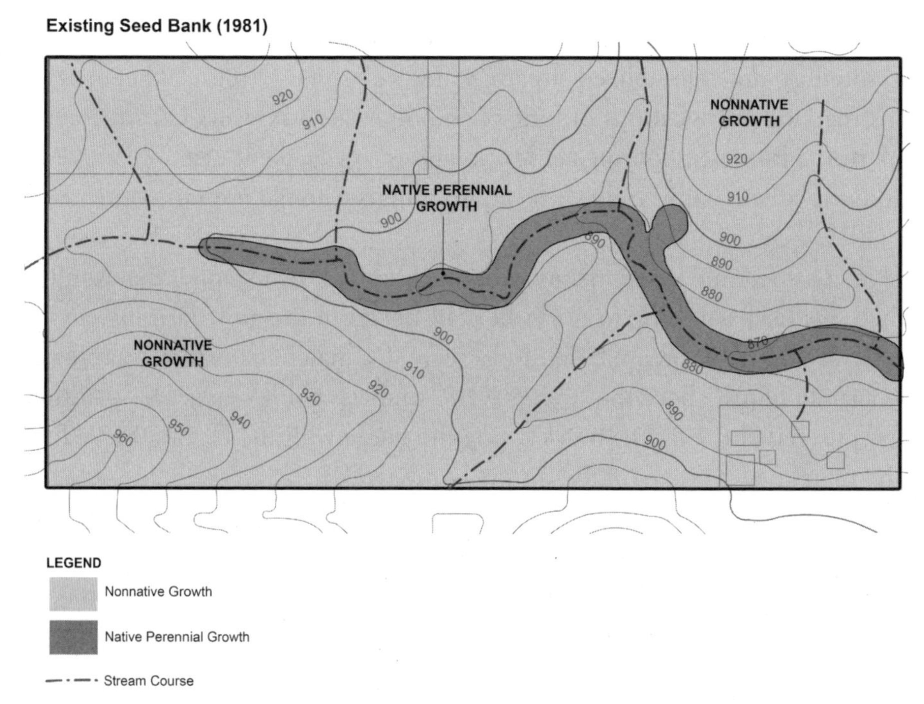


**FIGURE 3.5** Existing seed bank resources present on Stone Prairie Farm, 1981.

stream also contained many weedy species, as well as some interesting native species, including blue vervain and smartweed, likely to emerge from these soils. Areas where native species were still represented in the seedbank were mapped and distinguished from areas where only weedy species remained. In addition, the stressors that were still acting to degrade the landscape were summarized to show where key changes in the land and ecosystems have occurred (fig. 3.6) and then further summarized to generally label the most degraded to the highest health conditions on the land (fig 3.7).

## *Step Two: Investigate the Landscape History*

In this step, historic documentation is compiled, including historic aerial photos, conversations with older neighbors, and historic documents that describe how the landscape once appeared and how it has changed. Depending on where your property occurs, historic information will vary. In the eastern United States, you may need to reach back over two centuries to find presettlement information. In the West, you may find information from only one hundred years or less. Soils nearly always are important for interpreting the past, because they integrate conditions over very long periods of time.

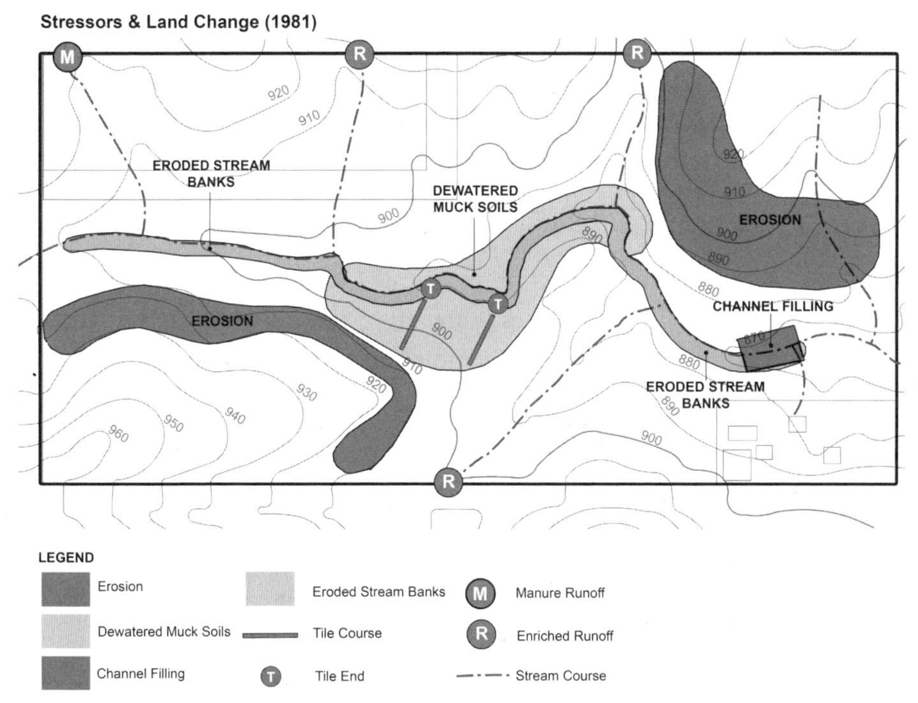


FIGURE 3.6 Stressor and land change summary at Stone Prairie Farm, 1981.

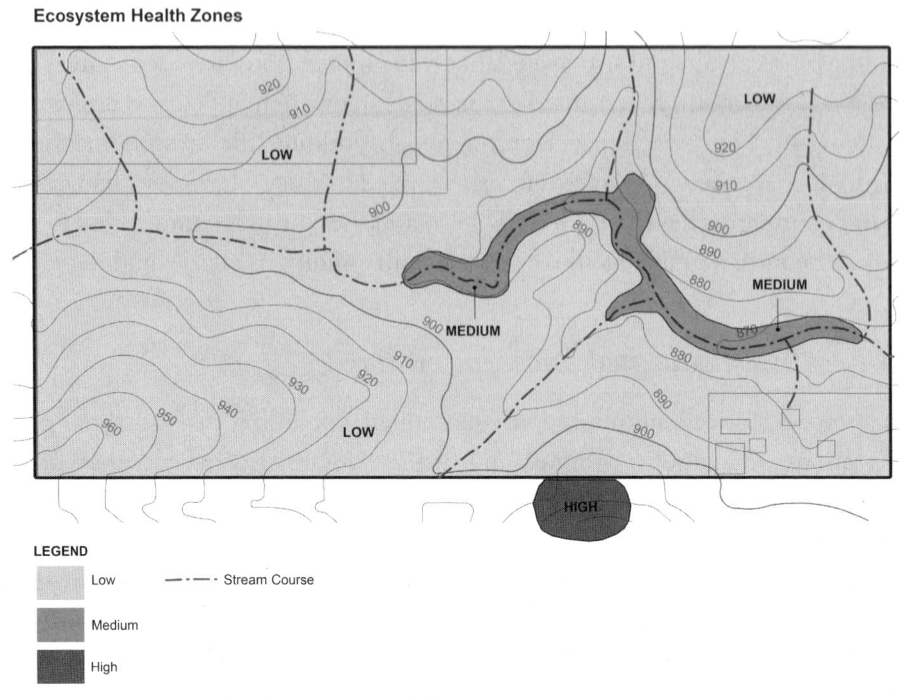


FIGURE 3.7 Ecosystem health zones in 1981.

At Stone Prairie, Steve and Susan began by mapping locations where soil types indicated that forests and prairie were present. They mapped the historic wetland locations based on the locations of existing springs and seeps, and the presence of the organic soils with very deep topsoil. Finally, based on stream terraces and upper boundaries of known flooding, they adjusted the historic wetland boundaries to fit the landforms of today.

If you are in an area that was primarily forested, check local soil survey publications to determine the species of trees that were primarily associated with different soil types found on your property. The county soil survey booklet will give some information on this. You can cross-reference to natural areas or other references to determine the nature of the forests that may have once occurred. We will go into more detail about these approaches in the following chapters.

For each of the ecosystem types that once occurred at Stone Prairie, Steve and Susan attempted to learn more about what they were like by visiting nearby preserves with the same ecosystem types and reference natural areas in the neighborhood where at least some aspects of each type could still be found, studying historic land survey notes and associated maps for clues as to how extensive the type was and which landforms it occurred on, analyzing older aerial photographs, having discussions with older neighbors, and studying the growth rings of older trees to get insight into climatic fluctuations or disturbance of the historic forests.

**Reference natural areas.** Steve and Susan located prairies, forests, savannas, and stream courses within a few miles and with similar features to Stone Prairie. They also obtained locations of state natural areas and other nearby preserves. If there is a natural areas inventory program in your state, it often is an excellent source of information. Steve and Susan were especially interested in areas for which they could obtain lists of plants and animal species. They matched the soil characteristics at Stone Prairie with similar soil in the reference areas then closely examined the vegetation in the natural area. When possible, they also reviewed information on the history and the management of reference areas.

They paid particular attention to the relationships of the reference ecosystems in regard to the landscape location, soils, and site history. Were the wet prairies and dry prairies associated with clear elevation or topographic features? Where did the forest or savanna edges occur relative to soil and landscape positions? Were there forests or savannas in lower elevations that would get seasonally flooded? They also were interested in how various plant communities blended with adjacent communities. Were boundaries abrupt or gradual? Where did communities occur along slope or hydrologic gradients, or disturbance gradients? Steve and Susan photographed examples of each community type, trying to illustrate their locations in the landscape and relationships to one another.

From visits to reference areas, they developed lists of native species that seemed to characterize each type of plant community. They did not attempt to compile exhaustive lists but included those species that were consistently present, especially those that were common. They then made a master list with a column for each plant community type that occurred at Stone Prairie. These locally derived species lists were used to determine the species and quantities of seeds or plants they needed to restore each community type on Stone Prairie Farm.

They also found remnants of some wetland and prairie communities on neighboring property near Stone Prairie (fig. 3.8). After obtaining permission from the private landowners, they explored those areas and added some additional species to their lists. They revisited these nearby examples many times, over several years, and often discovered additional species not previously seen. They also mapped the locations of the remnant communities on the most recent aerial photographs and on topographic maps. This allowed them to better understand relationships of each type of ecosystem to topography, hydrology, and soils.

Steve and Susan then extended their composite map of nearby remnant plant communities to create a map showing the historic distribution of community types in

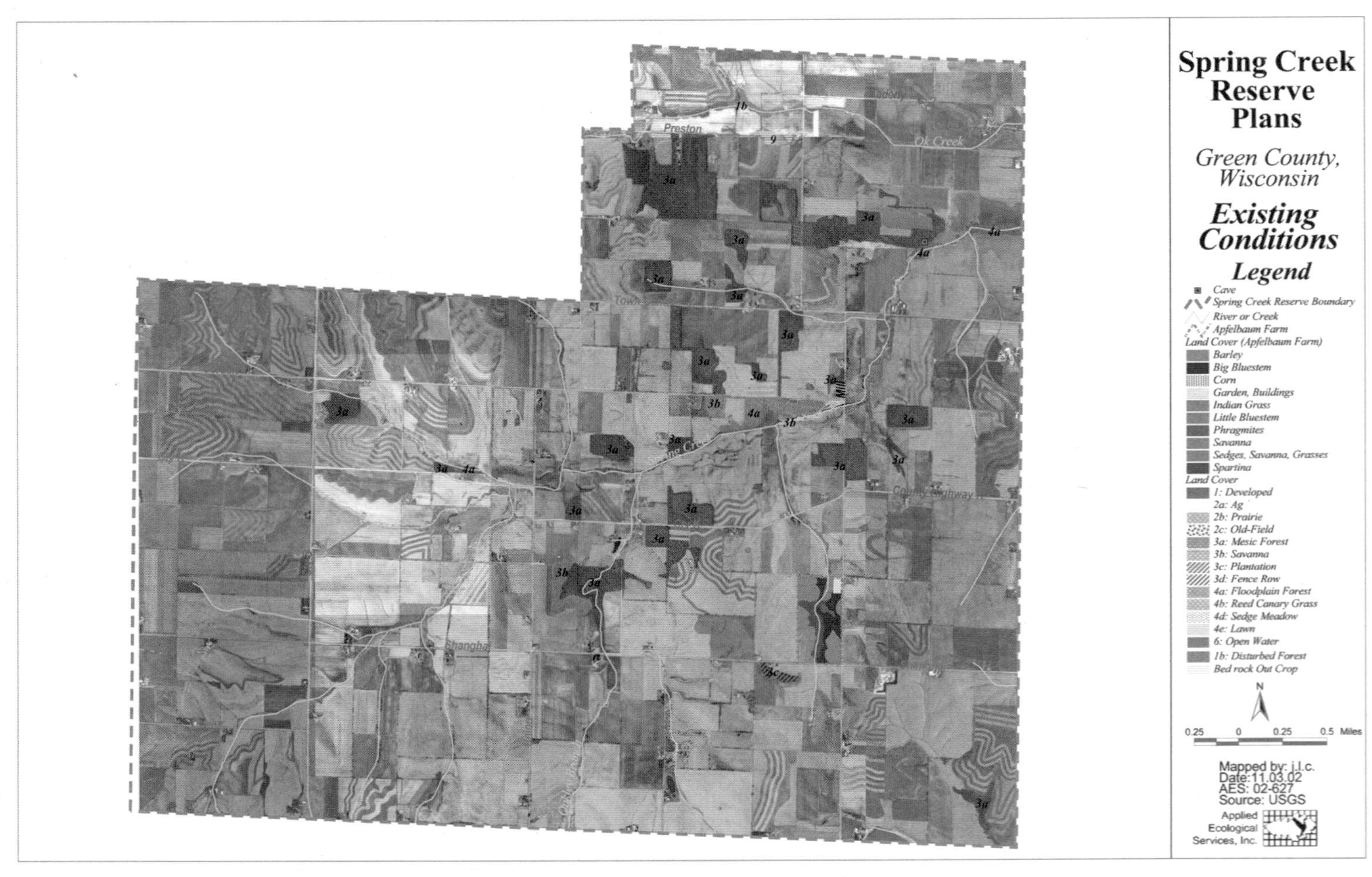


FIGURE 3.8 Natural areas inventory of the Stone Prairie neighborhood.

the wider neighborhood. They did this originally by hand, tracing community locations over local soil maps and on a topographic map. Then using the locations of each ecosystem remnant, they drew the extension of each remnant with colored pencils, using landscape locations and soil types, to create a picture of what probably was the extent of the historic ecological systems across and around Stone Prairie. Many years later, they did a more complete study using GIS (fig. 3.9).

By studying their plant lists and notes compiled from the reference natural areas, Steve and Susan could estimate the abundance of each species in each community type. They ranked them on a 5-point scale based on cover in the plant communities in the natural areas and in local remnants (5 had >50% cover, 4 had 40%–30% cover, 3 had 30%–20% cover, 2 had 20%–10% cover, and 1 had < 10% cover). The 5-point scale was developed to indicate which species dominated and, therefore, relative proportions to establish in restoring Stone Prairie Farm.

They also made photographs and sketches showing the interrelationship between different plant communities and their landscape positions. These sketches, together with the map of ecosystem types in their neighborhood, provided the basis for understanding the community types and their locations at Stone Prairie. This was the framework for their restoration plan.

**Original land survey review.** Stone Prairie Farm is in section 29 of Spring Grove Township, which is the six mile by six mile area where each square mile is numbered 1 through 36. The section line walked by the surveyors runs along the north-south property line. By reviewing notes for adjacent sections and overlaying observed features onto recent aerial photographs, they began to get a vision of what the surveyors found (fig. 3.10). They summarized the most pertinent details, which provided the following insights to the history of Stone Prairie:

- The uppermost tributaries of Spring Creek, including the entire south half of section 29, were originally prairie with a scattering of trees and interspersed wetlands, streams, springs, and spring brooks.
- The prairies occupied ridges and slopes, and, curiously, a narrow isthmus of prairie wove through the wetter valley and the side slopes that were otherwise forested to the north, west, and east of section 29. The southern side of section 29 had an expansive prairie.
- Southeast of section 29 was a large grove of oak, maples, and basswood surrounded by prairie. Other trees and their locations along the survey lines were identified by diameters. From this information, they estimated the relative abundance of species and their density.

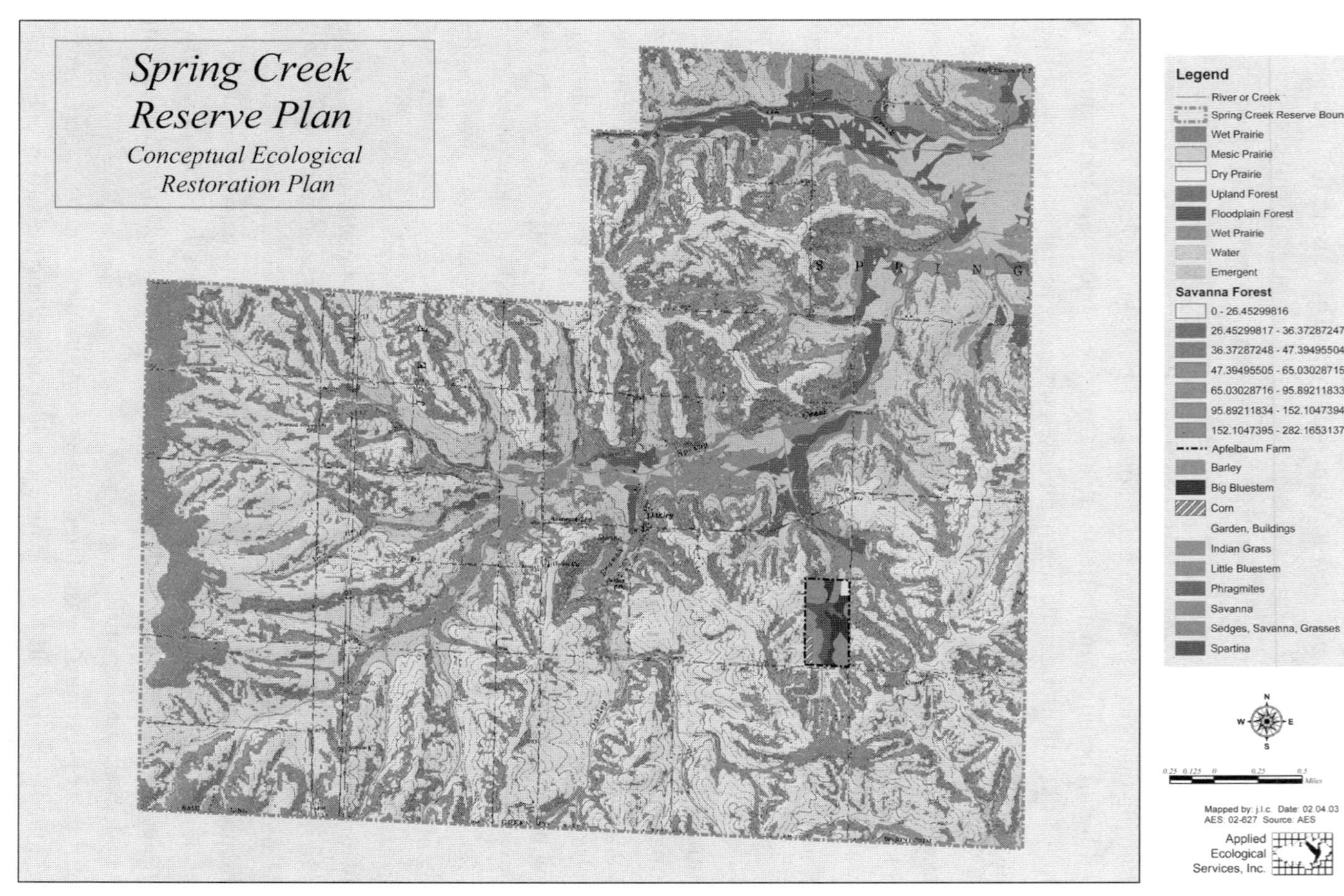


**Figure 3.9** Projection of ecosystem types that used to occupy the Stone Prairie neighborhood, based on locations of existing natural area remnants.

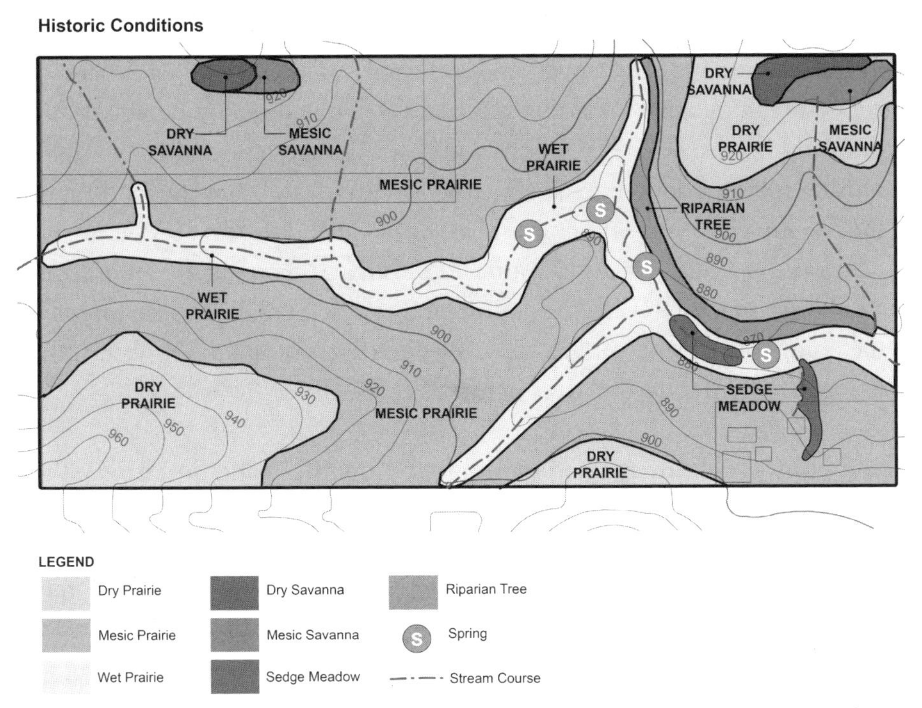


FIGURE 3.10 Historic conditions: summary of original land survey for restoration planning purposes for Stone Prairie Farm.

- North of section 29, a marsh, forested bottomland, springs, and brooks were found for nearly four miles.

The important conclusions regarding the original survey, formative in the development and later implementation of the restoration plans for Stone Prairie, were the following:

- Stone Prairie Farm was found at the interface of several major ecological systems: prairie, forest, wetland, and savanna, with considerable variation depending on location within the landscape and climatic changes over years to decades.
- Because of the mixing of ecological systems within a relatively small area, section 29 would have been a highly diverse and dynamic landscape. During years of unusually low regional moisture, wildfires would have been more frequent, larger, and probably more intense. Forest cover would be pushed back and opened, and prairies would expand. Oak-dominated forests might be converted to brushy oak woods, like those identified in the original land surveys.

Also, understory shrubs and sapling ingrowth would have been thinned in forests that retained tree cover, and some oak savanna would have been converted to prairie with oak grubs, a prime sharp-tailed grouse habitat. Indeed, prairie chickens and sharp-tails probably would have coexisted.

- During moist periods, trees such as aspen, American basswood, and American hornbeam; vines such as grape and bittersweet; and shrubs of American hazel, dogwoods, nannyberry, American wild plum, and prairie crab, among others, would become established in prairies and prairie openings in the forests. In the continued absence of fire, these species gave rise to thickets, brush prairies, and thick shrub layers in forests.

**Aerial photograph interpretation and mapping.** Steve and Susan were then eager to learn how the landscape had changed, leading to what they could now see. From the local NRCS office, they obtained copies of some of the cropping history and air photos used to check acreages. They obtained copies of images taken during the growing seasons of 1937, 1940, 1963, 1966, 1969, 1976, 1980, 1986, 1999–2001, and 2004. Each was roughly centered on section 29 and included surrounding land for a mile or more.

Steve and Susan made hand tracings (predigital era) of features they could identify in each photograph and categorized the changes from year to year. Anomalies observed were better understood by walking the land. They discovered an old bridge foundation over the creek, and the location of an old water well covered with soil. Erosion gullies showed clearly on the earlier photographs, and in the field they were able to confirm the gullies and observe how they had changed over time. They also understood the cropping history, and were able to calculate and classify erosion risk zones. They could confirm where the topsoil was lost and where it went.

When completed, this provided a graphic record of many of the features and changes on the land over time. Some information, such as seeing the locations of farm drain tiles when they were installed in the 1950s, was especially useful. They also could see exactly how the stream course had been altered, and how, in turn, that changed the hydrology of associated wetlands. These observations became key to understanding the importance of undoing those alterations during the implementation phase of restoration.

In summary, the following changes were concluded from studying historic aerial photos:

- Most land had been cleared and was being cropped in all of section 29 and across much of the township by 1937. Three years later, the lower reaches of the stream draining section 29 had been channelized, but upper reaches of Spring Creek were not. Wetlands along streams were smaller and had been partly drained and were being pastured, as evidenced by cattle trails through the vegetation. Bottomland timber was reduced to widely scattered large-crowned trees, probably bur oaks. At Stone Prairie Farm, the creek did not have a clearly discernable channel, which suggested that the channel elevation was close to the field elevation. Drainageways supported a loose scattering of trees and also appeared to be covered with riparian grasses or other vegetation.

- By 1963, the upper reaches of Spring Creek were channelized and shadows suggested that down-cutting had occurred. Tree cover had increased along the stream. Beyond the west fence line, the drainageway supported continuous tree cover of large-crowned willows, and perhaps elms. Old fencerows were conspicuous with scattered trees and shrubs. New fields were bordered by fences without woody vegetation.
- By 1969, the floodplain along Spring Creek was planted to corn, and many of the larger bur oaks and other trees had been removed. Some of the historic oxbows that were abandoned when the stream was channelized appeared to be widening. It appeared that a series of tile lines had been installed at Stone Prairie Farm to drain the southwestern-most field.
- By 1976, the abandoned oxbows along Spring Creek were obliterated by farming activity. North-south oriented linear features were probably recently installed tile lines. Tributaries to the stream draining section 29 had all been stripped of tree cover. Constructed sod waterways were apparent. In addition, the central drainageway had been regraded, fenced as a narrow pasture area, and served as a sod waterway. A uniform fine texture in the photograph suggested grain farming occupied the tilled land.
- By 1980, contour- and strip-cropping were present in adjoining farms, although not on Stone Prairie Farm. Spring Creek erosion was continuing. A young pine plantation bordering the farm was conspicuous. On Stone Prairie Farm, all but thirteen acres around the house were planted in corn.
- Steve moved to Stone Prairie Farm in 1981 and, thereafter, was able to correlate more information with the aerial photographs.
- The year of 1986 was very wet. Wetland soils were clearly identifiable on aerial photographs as dark splotches. Upland drainageways looked like varicose veins, and tiles were clearly indicated by narrow lines of lighter colored soils through the wetter, dark splotches. Corn was king for the last time at Stone Prairie. Some differentiation of upland grassed areas and grazed wetlands in the fenced riparian corridor were clear.

- From 1992 through 2007, photos revealed both dry and wet years. Especially during wet years, many historic wetlands stood out as very dark areas. The linear tile patterns showed up well. Prairie restoration was under way and results were conspicuous for the first time in 1999. In all fields except the northwestern-most, Stone Prairie Farm appears different from surrounding farms. The restored stream channel could be seen as a meandering dark thread, in contrast to the straight-line eroding channel a decade previously. A restored wetland that was filled with water showed up as a comma-shaped feature near the geographic center of the farm.

**Oral history interviews.** There were additional questions that had not been answered by the above analyses. What became of the dredge spoils when Spring Creek was channelized? Were there flooding problems after the land was cleared? Steve also wondered when elk, timber wolves, black bear, or prairie chickens were last seen in section 29.

Because Steve and Susan had not found this information in any documents, they needed to locate some local folks who might remember the early years. This began the final quest for filling in the gaps in some historical understanding. Some of what they wanted to know had little to do with what they could hope to accomplish through restoration at Stone Prairie Farm, but they were convinced then, as they are now, that this larger, richer understanding would give them a better sense of the historic ecosystem and how it once functioned.

They located some folks who had been intimately associated with the land. Some had been trappers and hunters then, if not full time, to supplement income and food on the table. They had interesting conversations with some arrowhead hunters who scoured the neighborhood for artifacts. It was exciting to watch the rising interest and energy emerge from the wrinkled faces and faded eyes as their probing questions took them back to their early lives. Interest in their childhood memories was clearly an elixir that increased their vitality.

To push toward the information Steve and Susan especially wanted, it was necessary to prime the conversations with questions, but listening with an open mind also was important. After dozens of interviews, a coherent story began to emerge. The story, however, emerged with much chaff that had to be winnowed. They heard stories of the wolf bounty hunters and unsolved murders. They listened patiently to stories about different family events—who married whom, and where they settled, and when their barns burned and their cattle were lost. They also heard myths perpetuated through three or more generations and stories of natural history that often confirmed some of their hypotheses.

Interviews were taped and the relevant information was transcribed. During the transcription, new questions and insights arose. Follow-up interviews were scheduled, which often were more helpful than the first. Steve and Susan learned more about the human relationships with the land than any aerial photograph or map could provide. They also realized that there would always be an endless and growing list of new questions to explore, but the source from which they gathered the social history was finite and shrinking.

**Tree ring analysis.** Steve used an increment borer to examine the history held within the few remaining bur oak trees around Stone Prairie Farm. Each year was nicely revealed along the pencil-sized core of wood removed by the borer. A ring of wood is laid down by the cambium layer each year, the oldest being at the center of the tree, and the youngest being just beneath the bark. By counting the rings Steve and Susan accurately estimated how old a tree was.

The trees ranged from 170 to 290 years of age. When the rings were closely spaced, it meant the trees were growing slowly. Slow growth might be associated with drought conditions or limited light because the tree was beneath a canopy of other trees. Widely spaced rings suggested years of good growth. Patterns of slow and rapid growth were found. There were approximately three times over the past 290 years where regional moisture was probably above normal for periods of 5 to 12 years each. These occurred at about 70- to 90-year intervals. Droughts occurred more regularly, but there had been three to five periods of intense drought, and they also appeared to last 5 to 7 years each.

Most interesting was the presence of scars on the lower trunk of many of the bur oaks, caused by wildfires that scorched the cambium, leaving a scar that the trees eventually healed over. Although not every fire leaves a record on surviving trees, most do, at least on one side near the base of some trees. The more intense fires leave records on nearly every tree, usually on the downwind side where fire wraps around the tree. Fire dates in the bur oaks corresponded well with drought periods.

## *Step Three: Interpretation of Landscape Changes*

The investigations of Stone Prairie Farm—tree ring analysis, soil mapping, conversations with neighbors, county tax records, property deed and title, land survey note—gave an incomplete picture of what Stone Prairie Farm and nearby areas once were, and how they came to be as they now appear. Like crime scene investigators, Steve and Susan searched for clues. Each clue provided a snapshot in time but began to fit into a complex story of how the land changed continually, like a kaleidoscope, leading to the present. Their goal was not the history so much as a desire to document with a high level of certainty what the historic ecological conditions were, summarized in figure 3.10. However, the understanding behind each mapped area was a complex story of how nature and humans interacted. A bit of these stories is provided here to illustrate the kind of information ideally needed to prepare a good restoration plan. What follows is an extrapolation of the information gained from those investigations.

As the land was settled and cleared for farming, widespread, often intense, wildfires became less common and, eventually, rare. More people led to more frequent fires that were smaller and less intense. Under this fire regimen, woody vegetation encroached on remaining prairies and savannas, even filling in open wetlands.

Most of the prairie sod in section 29 was converted to agriculture in less than two decades, and remaining wetland and prairie remnants declined as woody vegetation expanded. Patches of residual forests were grazed and cut for firewood and timber. The ecotones became sharper as agriculture expanded to the edges of forests, wetlands, and residual prairies.

Then German tiling and ditching technologies were imported. Starting at the spring brooks and creeks, drainages were deepened by dredges. From the ditched streams and wetlands, lateral ditches were extended to upland transition zones. This dewatered the entire wetland system. A second wave followed when clay tiles were installed, creating a lattice of additional drainage through wetland soils. This also reduced the storm water retention capability of the ecosystem. Dewatered bottomlands were converted from grazing to cultivation.

In uplands, forest groves were reduced to smaller and more isolated patches. Prairie largely disappeared from the landscape, relegated to isolated irregular patches of land ownership and land along streams and rocky ridges that would not support crops, where soils were too shallow to plow—at least more than once. As topsoil eroded, filling in the lower ground, much of the seedbank was depleted, and ridges and slopes lost the fertile soil that had accumulated over more than ten thousand years.

One of more interesting features in section 29 was the mosaic of mesic forests along Spring Creek, where hophornbeam, basswood, black maple, and northern red oak grew as outlier groves within a matrix of prairie, savannas, and dry mesic forest fringes dominated by bur oak and white oak. The prairies and small wetlands continually received seeds dispersed by wind and animals that crossed between these ecosystems. Berries, hazels, moonseed, and catbriar from the bur oak savannas moved across from the outlier groves and fringing bur oak forests. Trilliums, blue hyssops, phlox, bluebells, American bladdernut, violets, and many other species dispersed from the mesic forests in the valley and moist side slopes. Fingers along the many cool springs and spring brooks, and east and north side slopes following the landscape, likely supported comingling of all three of these vegetation types.

What a splendid place section 29 must have been! And what a magnificent location to live and gather foodstuffs! A trail system extended from the Sugar River up Spring Creek valley and appeared to end in the upper marshes and forest patches, suggesting the importance of this area for seasonal food gathering. Marshes and narrow, perennial spring-fed brooks would have been a haven for early-spring harvesting of common suckers and northern pike. Imagine the waterfowl use in pockets of wetlands and long, linear, east-west wetlands protected from early winter winds out of the northwest. Imagine the late fall holdover of waterfowl, as this valley may have been one of the last local areas to freeze, and may have seldom frozen completely as a result of the huge number and wide distribution of springs present. Waterfowl likely remained in the valley until late into the fall and provided a reliable harvest of meat. Good water, with prairie and wetlands interspersed with forests, would have been prime habitat for elk. The spring brook that runs through section 29 was likely one of the major north-south corridors for elk between the Spring Creek valley and the prairie groves to the south.

If you have been thorough in gathering information as suggested in the previous two steps, you will be able to construct an interpretation as Steve and Susan did. Keep in mind that any interpretation must be largely conjecture and hypotheses. More insight into how well you have interpreted the landscape will be obtained as you begin to initiate restoration and observe responses.

**Stressors and land change.** With a new piece of tracing paper, draw summaries of the existing stressors that appear to be deleteriously affecting the current and future restoration conditions. You will want to extend your interpretation to surrounding property, as many stressors originate off-site. Stressors include erosion, hydrologic alterations, landform changes (e.g., channelization, filling, burial, etc.), and fertilizer and animal-waste runoff. At Stone Prairie, stressors included areas where sedimentation and surface and ground water diversion had altered the hydrology and chemistry of water in springs and streams. Be aware of locations receiving highway runoff, septic-field water, farm fertilizer, and manure enrichment. These can be briefly described and shaded to indicate areas most severely impacted (fig. 3.6).

Biological stressors include locations of disturbances that either are currently supporting nonnative and invasive plant species, or are thought to provide opportunities for these plants to become established. At Stone Prairie, this included the muck soils dewatered by tiles, eroding stream banks, and locations where runoff from neighboring farms carried animal waste onto the property.

**Ecological health zones.** Based on all the information you have compiled, on another piece of tracing paper identify zones on the land that exhibit good, medium, and poor ecological health (fig. 3.7). Healthy zones should have few identifiable stressors; a medium condition is where minor stressors are present and often an abundance of nonnative species; poor ecological health is indicated in areas with several, often severe, stressors and few if any native species. This classification is intended to spatially summarize the condition of the land, and will guide the priorities and treatments required for restoration. The areas with greatest ecological health can usually be passively restored, often simply by removing livestock and by administering occasional prescribed burning, perhaps some spot treatment for noxious weeds. Areas with poorest health often require major intervention, such as noxious-weed removal at a larger scale, perhaps over several years before another introduction of native species can occur. At Stone Prairie, poor ecological health was found in areas such as spoil piles from the former dredging of the creek and old farm dumps and borrow-pits where gravel or soil was removed for use elsewhere. The value of identifying these areas is to provide a clear sense of the acreages and effort required to undertake restoration. It also graphically shows how remnant fragments can be reassembled to create more intact and continuous areas of restored lands. At Stone Prairie, this mapping identified small microsites with high native diversity embedded within other prevailing conditions. For example, a small gravel quarry with many native prairie plants was ranked higher, while the large ridge of which it was a part was found to be highly eroded and dominated by weedy plants. Consequently, it was ranked lower.

Gathering historic information and formulating preliminary hypotheses of what the land was like, how its ecosystems functioned, and the changes that have occurred over time is a rewarding exercise. Translating this information into a clear vision to guide restoration planning and future decisions is the creative work. This is the next focus of restoration planning.

To guide you, there are two key questions to ask: What were the historic ecosystems on the property, and what historic conditions can and must be restored? To answer the first question, a reference time period is useful but not necessary. Availability of information may limit your choices. At Stone Prairie Farm, Steve and Susan chose the earliest clear records, approximately 1837, when farming was beginning in the area. Write answers to these questions in your restoration notebook. To illustrate the level of detail desired, Steve and Susan developed some answers. The historic ecosystems present at Stone Prairie Farm were thought to be the following:

**Primary stream.** The primary stream originated in the upper reaches of the property where three branches converged. The edges through much of the property were lined with hundreds of springs. The upper reaches were fed by a very shallow sheet flow, along a gentle gradient, through prairie, savanna, or forest vegetation. An indiscernible channel meandered through wet prairie and sedge meadow wetlands. Each of the three tributary branches originated from springs that emerged from the toe of the bank. Water would have been clear and cold, flowing slowly in narrow, shallow rivulets, less then two feet wide and two to ten inches deep. Rivulets meandered through narrow streamside sedge and prairie grass stands. In the main branch, a slightly faster, narrower, deeper, cobble- and gravel-bottomed channel was present. This channel meandered through the lower areas of our property. In the neighborhood, we found several excellent reference areas where a spring-stream-wetland system was still present. The wetlands occupied the muck soils; the stream traversed clay and silt hydric soils still present in the land.

**Springs.** A vegetation gradient existed, and certain plant species were found where cold groundwater welled up. Within a few feet, as it warmed, other plants and animals (e.g., macroinvertebrates) would have flourished. Larger springs would have served as winter refugia, where the moving water kept headwater streams from freezing. Deeper pools would have supported native fishes—rainbow darters, southern red-bellied daces, black-nosed daces, and as the water warmed downstream, brook sticklebacks, fan-tailed darters, even brook trout, would have found refuge year-round. Caddisflies and mayflies were found in the sandy and rocky cobble substrates around the springs.

**Riparian tree margins.** Scattered along the entire main channel and western tributary, on slightly elevated gravel and sandy soils, were black willows, occasional basswood, red and American elm, and a scattering of bur and swamp white oaks. These trees created a shaded microsite that shrubs such as American hazel, American black currant, common elderberry, raspberries, fox grape, and numerous other plants colonized. Many of these same plants grew in the savanna copses.

**Dry prairies.** The highest gravel and thin-soil areas on the bedrock ridge were occupied by dry prairies, dominated by short-grass assemblages. Dominants included little bluestem and side oats gramma as the matrix, with northern drop seed on the north slopes, and seventy to two hundred other native species distributed along slope-aspect and moisture gradients. On the driest south and west aspects of the hills and ridge were lead plant, pale purple coneflower, pasqueflower, and porcupine grass. On northern aspects were shooting star, yellow lousewort, and birdsfoot violet. Soils harboring these assemblages were gravel and sand, or broken fragmented limestone with weathered clays over bedrock.

**Savanna copses.** Scattered across dry ridges were bur oak, black oak, basswood, occasional elm, butternut, eastern red cedar, and several other tree species that were able to become established in the prairie matrix. Beneath and around the trees, and even extending beyond their reach, were such native shrubs as American hazel, grey dogwood, hawthorn, nannyberry, sumac, elderberry, and a tangle of vines: fox grape, bittersweet, moonseed, Virginia creeper, to name a few. These copses of trees were irregular and highly variable in form, appearance, and size, and were the most dynamic and changing feature on the landscape. During heavy wildfire years, they would diminish, and standing dead snags would be observed, beneath which grew suckers from the roots and fire-scarred stumps of these same trees. They were a haven for many of the birds, mammals, reptiles, and other creatures of the prairie ecosystem. Most of the plants in these copses had bird- or mammal-disseminated fruit and seeds (e.g., berries, grapes, haws, rosehips, stick-tights, nuts, etc.), and the movement and development of copses depended on blue jays, raccoon, fox, opossum, elk, white-tailed deer, and other species to disperse these propagules.

**Mesic prairies.** This ecosystem developed on well-drained but seasonally wet soils comprising clay and silt loams with deep, well-developed, dark topsoil. The dominant plant community was characterized by tall native grasses—big bluestem and Indian grass—with several hundred other species of plants, including rosinweed, compass plant, bergamot, mountain mint, sunflowers, and golden alexander. Many species were associated with subtle changes in soil, topography, or hydrology. Over areas with a persistent, higher, springwater table, species such as prairie cordgrass, dozens of sedges, and many wildflowers, such as mountain mint and sweet black-eyed Susan, were dominant.

**Wet prairies and sedge meadows.** These areas were wet through midsummer during a normal year. Soils were a mix of interbedded shallow muck—dark, organic soils—and layers of very black silt and clay subsoil, mottled with streaks of gray indicative of reduced iron. The upper, drier margins of these wet areas held many of the same species as wetter areas of mesic prairie. Additionally, species such as cup plant and golden glow, sand bar willow, meadowsweet, and steeplebush were common in this zone. Also present were prairie cordgrass swards that irregularly interdigitated with adjacent mesic prairie. In persistently wetter areas, along the lower margins, especially bordering the stream, dense growths of red-node sedge characterized this ecosystem type.

For all of these ecosystems, fire was important. During dry periods, usually in the spring, fire would sweep across the landscape. It most frequently burned the upland communities, the dry prairie, and especially savannas that held more fuel. Fire often burned the mesic prairie, although sometimes the fire was light or skipped much of it. It sometimes burned through wet prairie and riparian communities, although it very seldom burned into the peat. Fire released accumulated nutrients and stimulated growth of nearly all species of plants.

Grazing and browsing animals such as elk, deer, and maybe bison were important, although insects consumed far more biomass. Nutrient cycles were enhanced as plant tissues were digested and combined with nitrogenous wastes, leading to greater plant productivity and better soil development. Erosion was nearly absent, even after fire or heavy rain. Plant roots formed a thick sod, and most species quickly recovered from fire through vegetative regeneration. Rain and snowmelt were absorbed into the ground, and springs ran freely through the summer, even through all but the most severe droughts.

Which historic conditions can and must be restored? This question is both a reality check on what is possible, and begs the question of what is meaningful to you. Critical to the answer is the level of effort and expense available to support the restoration. Historic conditions, realistically, can never be achieved, if for no other reason than extirpation or extinction of some of the species once present. Moreover, historic fire and grazing regimes can no longer be duplicated. The questions thus become what realistic level of restoration is desired and are the available resources sufficient to get there.

Approach these questions by referring to the map showing ecological health zones (fig. 3.7). For Steve and Susan, the immensity of the restoration effort required to reach their goals at Stone Prairie became very clear when they reviewed their map, along with the map of the stressors that would have to be addressed.

Some of the stressors were ongoing (e.g., nutrient enrichment from neighboring farmlands and feedlots, hydrology alterations in tributary lands, and presence of persistent invasive plants and animals. It was impossible to achieve the historic conditions they naively hoped for when they began. In discussing their frustration with a colleague, he offered this counsel: "Don't become a slave to the past."

The inability to duplicate how historic ecosystems looked or functioned does not mean that you cannot achieve important and meaningful ecological restoration. The complex diversity of healthy ecosystems provides much duplication of form and function, such that ecosystems usually perform remarkably well even when diversity has been compromised. The most important function to support is *succession,* defined as an orderly replacement of species over time, gradually changing the chemical and/or physical environment, favoring the next suite of species. A more pragmatic view of succession is as an ecological repair process. It is initiated by disturbance, and the sequence of species, by and large, restores the structure and functions of the ecosystem. It is a never-ending process, and it also can be thought of as an ecosystem maintenance process. It is not necessary to restore all the structure and functions of the ecosystem. The aim is to assist natural succession in restoring the ecosystem to a level where it can continue to maintain itself.[1] So how do we jump-start succession?

Jump-starting succession involves two primary strategies: reducing or removing the major stressors that led to degradation of the ecosystem; and introducing as much of the native diversity as is feasible, especially those species and associations that are keystone contributors to structure and functions.

## *Step Four: Develop Realistic Goals and Objectives*

This task is one of the most pivotal in the restoration-planning process. It is now that you decide what is realistically restorable, given limitation of resources and minimal requirements of the ecosystem. What cannot currently be restored also should be understood and documented.

*Goals* are the overarching statement of what you are trying to accomplish. For example, a goal might be restoring stable soils or native diversity that equals the condition in a reference area. *Objectives* are what you actually wish to see or measure as changes on the land. A goal of stable soils, for example, might lead to an objective of no measurable suspended soil load in water draining from the ecosystem. Qualitatively, the cessation of rill and gulley erosion from slopes also would indicate stable soils. As much as possible, objectives should be worded so that they can be quantitatively evaluated.

For Stone Prairie, Steve and Susan considered the following questions to come up with their goals and objectives:

- What cannot be restored because of significant landscape changes or extinction of keystone species?
- What cannot be restored because of costs?
- What is not likely to be restorable without neighbor cooperation because of what neighbors do on their land?
- How do scale considerations affect both what is likely and the benefits of what can be restored on the land?

Steve and Susan decided that an overarching goal for Stone Prairie Farm was to become a regional refugium for the propagation and survival of locally declining rare species and the communities they need for survival and well-being. That established a benchmark for all of the goals that were then seen as specific conditions necessary to achieve that objective. It would be fairly easy to develop specific objectives to go with each of the following goals:

1. Stable soils.
   a. Stability of stream bank and bed within three years.
   b. Cessation of rill and gulley erosion from each ridge and farm field within two years.

2. Plant communities dominated by native plant species, with representative diversity and structure.
    a. Dominance of each restored ecosystem by a minimum of fifteen native plant species within five years.
    b. Diversification of each ecosystem with additional appropriate native plant species within ten years.
    c. Successful establishment and regeneration of at least twenty rare, special concern, or federal or state threatened or endangered plant species within ten years.
3. Communities dominated by native faunal species, with representative diversity and structure.
    a. Dominance of each restored ecosystem by native insect, mammal, bird, reptile, and amphibian species within ten years.
    b. Diversification of each ecosystem with additional appropriate native insect, mammal, bird, reptile, and amphibian species within fifteen years.
    c. Successful establishment and growth of at least ten rare, special concern, or federal or state threatened or endangered animal species within ten years. The objective is to have Stone Prairie Farm contribute to the regional refugia for propagation and survival of declining rare species.
4. Clean, cold water streams that harbor native fish and macroinvertebrates.
    a. Dominance of guilds and species of macroinvertebrates, fishes, and aquatic plants indicative of cold, mineral-rich springheads and headwater streams within five years.
    b. Diversification of each biological stratum with appropriate native insect, mammal, bird, reptile, and amphibian species within ten years.
    c. Successful establishment and growth of at least five rare, special concern, or federal or state threatened or endangered animal species within ten years.
5. Appropriate levels of run-on and runoff water, entering and departing the property and flowing into the stream.
    a. Biocleansing of run-on waters through restored wetlands within five years.
    b. Cleansing of storm water runoff from our farm yard within five years by biofiltering through grassed and restored wetland areas. Remove the straight pipe from barnyard to stream within three years.
6. Community and neighborhood connections to ecology through our land.
    a. Use of Stone Prairie Farm for educational activities by local school groups, neighbors, and other interested local citizens within three years, up to two groups per year.

   b. Use by persons from nonlocal school groups, regional, national, and international locations within three years, up to two groups per year.
7. Learning and sharing knowledge.
   a. A personal commitment of time and hands-on involvement in each restoration and monitoring activity.
   b. Document changing conditions in the ecology of the land and publish the story in popular press and technical publications within twenty-five years.
   c. Tell the story of ecological restoration as frequently as possible throughout the process.
8. What we can afford do to within our means.
   a. Maintain good records of our expenditures on restoration.
   b. Form partnerships with organizations and neighbors to extend our means, so as to be both more effective and expansively successful in our restoration efforts.
9. Through our restoration, contribute to the well-being of the Earth and human culture and ecology.
   a. Demonstrate how landowners of small parcels can contribute to larger outcomes through restoration.
   b. Demonstrate benefits to the local economy and culture.
   c. Document partnerships that have organically developed to extend the benefits from our restoration to other lands.
10. Passion for a healthy Earth ethic.
   a. Do two or more speaking engagements annually to spread the word.
   b. Document the response of others to the experience we have shared.

Goals and objectives are neither static nor developed in isolation. Both the temporal context and the knowledge used to formulate the planning change. Whenever it becomes apparent that something significant has changed, or important new information has been acquired, you need to refine or add to your goals and objectives. This is an application of adaptive management. At a minimum, ongoing restoration projects should be reviewed annually in a systematic way, and changed or adjusted where necessary.

### *Step Five: Prepare a Plan*

Gather all of the maps and tracings made through the previous tasks. Use the descriptions of historic ecosystems, your goals and objectives, and the existing conditions to refine your planning.

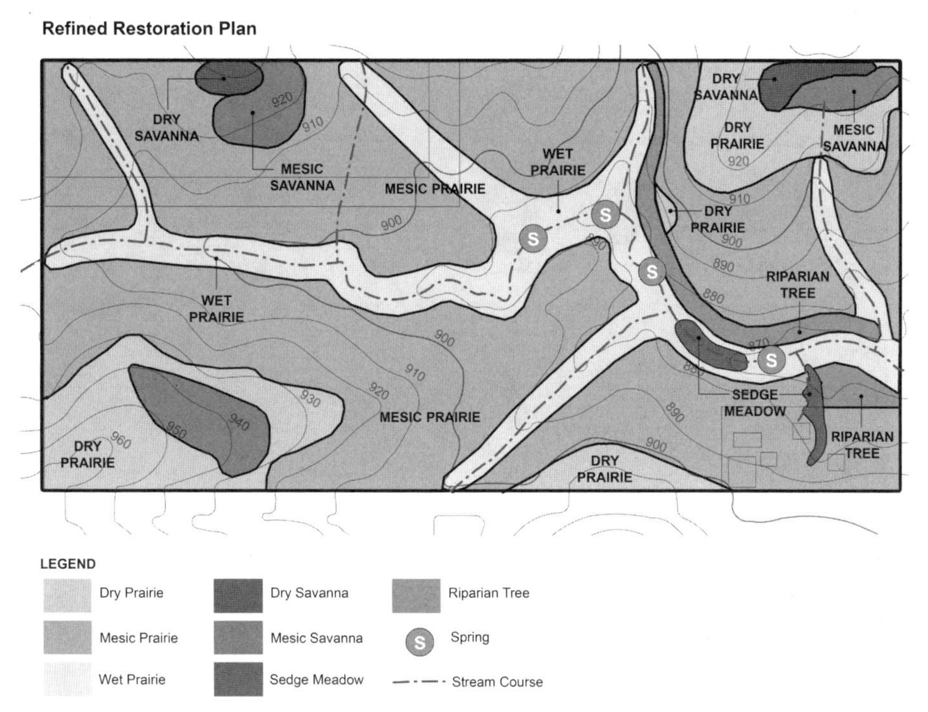


FIGURE 3.11 Refined restoration plan for Stone Prairie Farm.

A refined restoration plan lays out the specific methods, equipment, seeds, and other materials needed to implement the plan for each management unit. First, refine the mapping of each targeted ecosystem type, clearly showing where each would be restored on the land (fig. 3.11).

**Definition of management units.** These are the locations where you can successfully implement restoration during any year, and where location-specific management tasks need to be implemented. Management units are typically the mapped locations where similar types of restoration treatments will be conducted on the land. For example, at Stone Prairie, restoration of the channelized stream was identified as a management unit. Steve planned to backfill the dredged channel and redirect the stream back into the former wetland that included the historic meandering stream channel. Both the existing channelized reach and the location where the redirection and restoration were to occur were identified as management units (fig. 3.12).

**Phasing and scheduling.** Once management units are assigned, begin dividing the overall project into annual components for implementation of work. With your priorities in mind, develop annual budgets, projections of material needs (seed, plants, herbicide, etc.), labor, and equipment for the work planned each year.

An annual component is primarily based on what you think you can accomplish

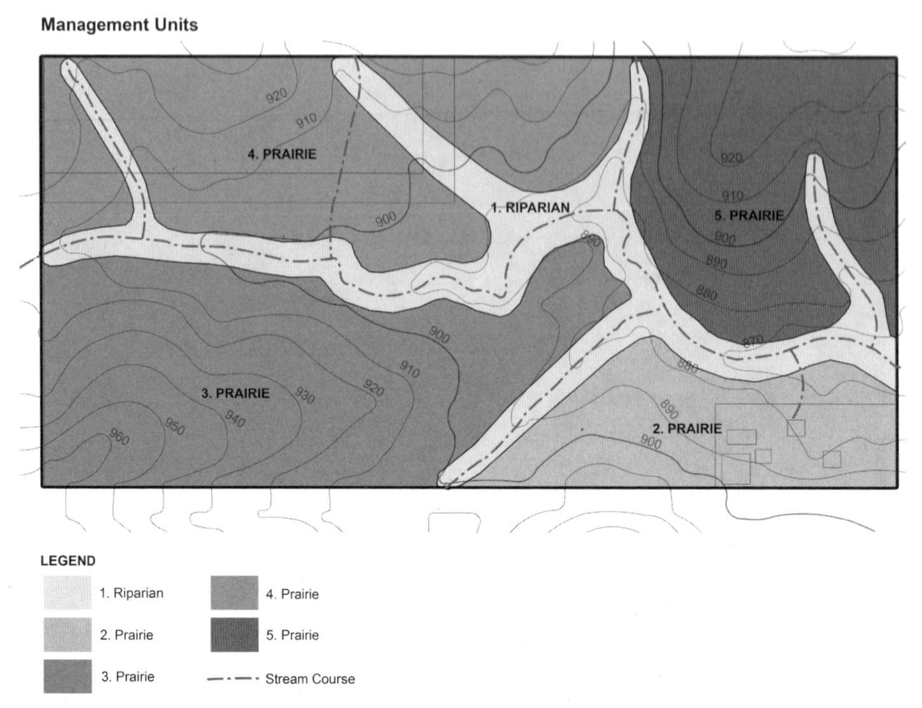


**FIGURE 3.12** Defined management units for implementing the restoration plan at Stone Prairie Farm.

each year. So while management units define differences in the restoration needs, annual plans reflect what you can actually accomplish (fig. 3.13).

**Methods and materials.** A critical step in restoration planning is defining the specific methods you will use. As with much in the plan, the more detail the better, but a minimum is essential for even rough estimates of cost, labor, and equipment. Once methods are defined, those components of each task for the year ahead should be itemized. It is a good idea, also, to identify the seasonal window for the operation. Timing for many restoration activities, such as burning, seeding, and weed control, is critical to their success. Even activities such as restructuring drainageways need to be coordinated with soil moisture and precipitation, particularly avoiding runoff.

Steve and Susan prepared a table with the management zones named or numbered in rows. The methods to be used were laid out in columns. They also identified the specific seed or planting mixes that corresponded with each management unit that was to be seeded. For each, they defined how they were going to conduct the seeding: broadcasting seed onto a plowed and disked seedbed; no-till drilling; or hand planting. In short, they developed a catalog that listed methods, materials, and equipment needed for undertaking restoration and management activities for each

FIGURE 3.13 Defined phasing plan for implementation of Stone Prairie Farm.

management unit for the coming year. Each year they expanded the catalog to capture the annual management plan. The catalog became a permanent record of activities. In time, the catalog was converted to computer spreadsheets, which makes the task much simpler. Depending on the complexity of your restoration, you may find this approach useful, or you may be able to develop a simplified way to catalog and track restoration activities. Regardless, take the time to keep careful notes of when and how things were done. Remember, much of ecological restoration is learning by doing.

**Schedule work.** Scheduling activities translates the preparation plan into an action plan, the what, where, when, and by whom. Note that in the goals Steve and Susan prepared a timeline was included. You likely will find that you need to revise these timelines, but the work schedule brings that thinking down to an annual work plan where your projections will be much more realistic. You now can schedule the labor, equipment, and materials needed for the months ahead. This also helps you project costs for cash flow purposes.

**Project schedules.** For any project of more than a backyard size, a schedule of work is usually warranted. Initially, Steve and Susan drew up a schedule on scratch paper, with rows and columns. Each column represented a month of the year, and rows were the specific tasks to be implemented. In right-hand columns, they listed the method to be used and who was going to actually implement each of the tasks. They did not have some of the equipment to do certain jobs. Also, they did not have the licensing or enough confidence in their own abilities for some jobs, such as herbicide application or prescribed burning. With a project schedule, they could arrange to hire professionals to do tasks for which they lacked equipment or expertise. With a project schedule, you can line up the help when you need it and make more precise cost estimates.

**Determining the labor and associated costs.** You need to assess your own monetary and time commitments to the restoration, and arrange for contracts or hiring as you can afford or need them. It is necessary to break down the annual plan into monetary and labor costs for each task. This also supports accurate cost projections and tracking of real costs (time and dollars).

Steve and Susan expanded the spreadsheet for each activity to include columns for the acreages in each management unit, and another column for unit costs estimated for each task (dollars per acre, dollars per hour). They then added a column for estimated hours for each task in each management unit. A final column was the multiplied values across all of these entries. This became relatively easy to do with spreadsheets, but it also was not that difficult with a hand calculator. Steve and Susan found it essential to have the annual summary taped to the refrigerator or somewhere prominent to keep the process always in the fore, as there are many distractions. With the reminder, they could then retrieve the detailed spreadsheet for the specifics.

This process allowed them to develop useful and increasingly accurate budgets over the years. With computer spreadsheets, making refinements in any element of this schedule and budget is easy.

## *Step Six: Develop and Initiate a Monitoring Program*

It is essential in the implementation of a restoration project to design and install a monitoring program. We provide details for monitoring in chapter 5. To the extent possible, monitoring should provide quantitative and qualitative data and photographs that allow evaluation of objectives and goals.

Ecological restoration should be viewed as neither conclusive nor absolute. Restoration and management programs must be flexible because of the unpredictability in ecosystems and the environment. The complexity of ecosystems makes it difficult to predict exactly how a system will respond to many, if not most, treatments. Programs need to be changed in response to new data and derived insights, or when treatments fail to provide the desired response. What we have covered so far is the beginning of an ongoing process of restoring ecosystem functionality, biodiversity, and natural processes. Regular monitoring during the restoration process will provide feedback on the program's effectiveness and generate information to evaluate and guide changes. This process of evaluation, adjustment, refinement, and change in the restoration program is called *adaptive management.* Adaptive management is fundamental to the restoration and maintenance of ecosystems.

## *Step Seven: Implement the Plan*

With the plan in hand and baseline monitoring completed, you can begin implementing restoration. Steve and Susan felt that when it actually happened, it seemed anticlimactic. They had been planning for two years, with only small patches of prairie around the house to whet their appetites for what was to come. Their first prescribed burn lasted less then fifteen minutes . . . two years of planning for fifteen minutes of action. While gratifying, it was a short-lived celebration. The real benefits from the hard work that had gone into planning were not to be realized or fully appreciated until two years after they started implementation, when they could see the results in the monitoring data or, for that matter, look out the window and see native prairie where Eurasian weeds previously dominated.

## *Step Eight: Document Changes and Maintain Records*

Recording where, when, and how restoration was implemented, and how the land responded, is very important. Following through on the monitoring and documenting the program results was one of the harder things for Steve and Susan because they were so busy trying to implement the plan. They purchased a large notebook and started a farm journal in which they wrote observations, and they made a regular practice of documenting questions, information, what they thought they were learning, and how the land was changing. They also captured information on a computer, with backup hard copy, including repeat photographs showing change, species lists, and diversity estimates, and anything else they could digitally use for documenting restoration work. These files were updated regularly, annually at a minimum.

Without the baseline and annual (or some alternative schedule, such as every three years) monitoring, it would be difficult to document details of the changes that occur on the land. You can remember the broad, sweeping changes, but the details are usually more critical. Understanding the rate of change, what species appeared or disappeared, what methods worked best and during which years (correlated with the meteorological records) is not only useful in planning for future restoration efforts, but also provides an accurate, ongoing assessment of how the restoration is progressing.

## Conclusions

The planning and implementation process described in this chapter was developed, tried, and improved during twenty-five years of restoration at Stone Prairie Farm and on hundreds of other ecosystems throughout North America. The before and after aerial photos from Stone Prairie Farm (figs. 3.14 and 3.15) attest to the validity of the methods described. In part 2, we examine how this basic approach can be applied in each major type of North American ecosystem.

In any complex restoration project, there will be considerable uncertainty. By ignoring the uncertainty, you risk greater cost and loss of time. The best way to mitigate uncertainty is to test hypotheses through experimentation. These may be small-scale demonstration plots, such as those adjacent to Steve's house, where he easily could keep an eye on them. When he felt he had gained enough certainty, the approach or methods were scaled up and applied to management units where appropriate.

We want to emphasize the inherent resiliency of ecosystems.[2] As long as the major stressors are correctly identified and alleviated, and as long as reasonable native diversity persists or can be reintroduced, the key compositional, structural, and functional characteristics found in ecosystems can, with your help, recover in time.

**2001 Land Cover**

**1940 Land Cover**

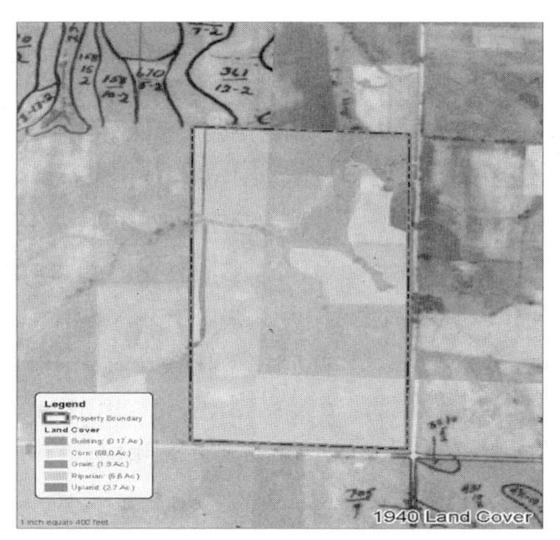


**FIGURE 3.14** A comparison of the land cover in 1940 and 2001 shows the large areas of cornfield that were converted to restored prairie vegetation. In 1940 the historic stream that has been restored and that is evident on the 2001 image was being completely farmed.

**October 2008**

FIGURE 3.15 The oblique aerial photograph of the Stone Prairie Farm taken in October 2008 showed the patchy habitat conditions restored, compared to several surrounding cornfields in adjacent farms. Restored wetlands and the stream course are also clearly visible in this image.

Much of restoration effort is to jump-start or speed up the process. Gaining confidence in nature's ability to heal itself allows you to focus on how to assist this effort. The belief that restoration is completely under your control is arrogant. Develop an understanding of the ecosystem, remove the causes of its degradation, and then let nature do most of the work. Planning guides this process, and requires an open and inquisitive mind, persistence, and humility. However, even with the most realistic vision and best strategies to accomplish it, someone has to schedule tasks at the proper times, make sure all necessary permits or licenses are obtained, line up necessary equipment and labor, and see that tasks are completed properly. This can only happen with good planning and follow-through, as described in the next chapter.

## Chapter 4

# Implementing Restoration

*The last word in ignorance is the man who says of an animal or plant, 'What good is it?' If the land mechanism as a whole is good, then every part is good, whether we understand it or not. If the biota, in the course of eons, has built something we like but do not understand, then who but a fool would discard seemingly useless parts? To keep every cog and wheel is the first precaution of intelligent tinkering.*

Aldo Leopold

So far in your planning, you have identified where you wish to go in your restoration. You now must decide how to get there. Having very clear ideas about what you wish for the land is an extremely important start, but restoration at any scale in all ecosystems requires money; labor; usually some materials such as seeds, plants, or herbicide; and equipment such as a tractor, disk, wick applicator, and sprayers. An early decision also is needed about priorities: where and how to begin. Given money, labor, materials, and equipment, time becomes the critical factor in determining a starting strategy. Time may not help you determine where to start but is very important to the scale of the operation you can manage. Timing, as opposed to the time available for working on a project, is also important.

Timing of soil preparation, planting, or invasive weed management often influences success as much or more than how you execute the task. For example, if you are starting from a raw patch of earth to reestablish a desert ecosystem in southern Arizona, both soil preparation and seeding should be done about the same time in the winter. On the other hand, in well-drained loess soil of northeastern Iowa, preparing a site for prairie often begins two years before planting. Preparation may involve early-, mid-, and late-growing season treatments, such as mowing, herbicide, or tillage, to reduce nonnative, persistent grasses before seeding or planting native species. Availability of equipment, materials, and labor can also influence the timing of when a restoration project can begin. Scheduling and coordination become important on even simple projects.

In this chapter, we describe how to decide where to start and how to schedule the work, and we review the general techniques and tools that are most commonly used. The aim is to increase your probability of success with a good starting strategy and proper techniques. Bear in mind that early project success will greatly influence the motivation to continue a multiyear restoration effort. Too many projects, even those that are well planned, are abandoned when initial efforts encounter unanticipated problems or produce results that fall short of expectations. We begin with some general considerations you should consider before beginning your project.

## Working at the Right Scale

Common start-up problems in ecological restoration can be characterized as (1) too big and idealistic or (2) too small and inefficient. The most important consideration regarding the size of your project is that whatever you start, be sure you have the capacity to follow through, including monitoring, to fully implement and maintain it. Until badly damaged ecosystems have regained considerable integrity, they require maintenance. Maintenance may be required for years in many ecosystems, although the amount of effort and time will likely decrease as restoration advances. Vigilance is required even for healthy ecosystems, especially in regard to invasive species and other *stressors* from outside. Starting too big without the follow-through can create more work, because treatments may fail and need to be redone. Until you gain some experience and comfort with the methods, starting smaller also allows you to get a better perspective on time commitments and financial requirements. Be careful, however, that a too modest project may quickly become boring, causing you and those helping you to lose interest. Volunteers, including family members, may not understand the grand vision or have enough experience to know that progress, especially in the first few years, can be painfully slow.

The best scale, we have learned, comes from understanding the types of treatments needed and the scales at which they most effectively and efficiently operate with the equipment available. For example, herbicide applications can be on almost any scale with the proper training, experience, and help; while tilling soil with a tractor and disk probably is not efficient except when addressing at least a few acres. Applying herbicide with a hand sprayer may make sense when treating areas smaller than half an acre, but commercial-sized sprayers are needed if you are working large areas. (We address the pros and cons of using herbicides later.)

Starting with small test plots is a good idea in many cases. If these are placed in locations with the most urgent restoration need, or used for the most challenging areas where efficacy of treatments needs to be confirmed, this strategy can help refine methods and build confidence as well as affirm material and labor needs. With effective monitoring of the practices and outcomes, you can test and refine methods before committing resources for a large-scale effort. Be sure, however, that the size of the test is sufficient to provide a good evaluation of the methods and treatments being applied.

When using prescribed fire to restore native plant communities, a common tendency is to start at too small a scale. Scale of fire can make a big difference. Usually, the smaller the treatment area, the more uniform the burn, which favors homogeneity in the responses. This is because the time period over which the treatment occurs is short, and humidity, wind, and fuel combustion conditions are more uniform. Consider how fires burned historically over the landscape, flaring up and burning hotter during the heat of the day, or on south- and west-facing slopes, and dying down at night, or in wetter pockets or north-facing slopes. Prescribed fire should encourage diversity of plant and animal communities rather than minimize them, and this is usually a factor of scale.

Avoid task-oriented strategies and focus on ecosystem-process strategies. This saves time and money and produces more effective results in both the short and long term. An ecosystem focus involves addressing the physical, biological, and chemical drivers or constraints that control the ecosystem at the scale they operated when the ecosystem was healthy. This means working with, rather than against, nature—for example, restoration of hydrology, altering pH, or restoring a disturbance regime. Typically, where these natural *drivers* have been altered, invasive species are prevalent. Invasive control is secondary to the underlying problem. If you ignore the driver, you may spend all your time fighting invasives.

The aim of an ecosystem approach is to trigger positive feedback loops, jump-starting the ecosystem on a trajectory of recovery. For example, by restoring water retention in soils that historically were wetlands, you disfavor invasive plants and animals and shift the advantage to native species, thus reducing the need for spraying or mowing. Hydrologic restoration, in a remarkable percentage of instances, is one of the most effective restoration strategies, with consistent benefits and few risks. Even in uplands, altered hydrology can have wide-ranging effects. For example, in an oak savanna restoration, Steve discovered that disabling tiles in nearby a field that was a historic wetland reduced invasion of European buckthorn. Likewise, filling drainage ditches can affect hydrology over a wide area, including neighboring property. Be sure, when addressing ecosystem-scale issues, that you consider off-site impacts.

Ecosystem-scale strategies may involve modifying human behavior or use of the property to accommodate seasonal changes. For example, in seasonally flooded agricultural lands, a strategy might involve allowing the farm fields to flood longer each fall or spring to provide temporary migratory waterfowl habitat, such as rice fields in the Sacramento Valley, California. Releasing water from dams to simulate annual streambed scouring and riparian flooding is another example of modifying use to restore ecosystem processes.

Ecosystem strategy sometimes exceeds ownership boundaries. Opportunity to operate at the proper scale may require neighbor cooperation, a special challenge if they do not share your vision and values. The aim is to address stressors or drivers at whatever scale they operate. For example, disabling a tile system of no more than a few hundred feet might restore a wetland, but controlling sediment and nutrients from upstream runoff would typically require cooperation of other landowners in the watershed.

If ecosystem processes are not the focus, you generally will be addressing symptoms. There are endless tasks that can appear useful if you only address symptoms, but which do not move you toward your goals. One of the greatest problems with small parcels of land is that it becomes difficult to work at the ecosystem scale. Even with smaller parcels, however, identify the ecological drivers and stay focused on them as much as possible.

## Where to Start

Unless there are special circumstances, such as rapidly spreading invasives or unusual opportunities, start with the healthiest areas: remnant forests, a tag of tallgrass prairie, or wetland vegetation hanging on along a stream margin. Look for some early successes and build from those refugia. In the case of Stone Prairie Farm, Steve and Susan chose multiple starting locations, recognizing that the native prairie remnants could become sources of seed for spreading this ecosystem into fallowed farm fields. They also realized that the hydrology restoration from breaking tiles and backfilling ditches would substantially change the ecology of large areas of the farm, so they did both. They also began cover-cropping a majority of the farmland to begin weed control and rebuild some impoverished and eroded soils. To generate some revenue, they rented some of the land to a neighboring farmer and required that he remove internal fences, thus accomplishing yet another goal in the first two years. In each project, there will be reasons to sequence work and tasks, or to place a higher priority on one thing over another.

## Working With, Not Against, Nature

We previously suggested the importance of working with nature and will continue to illustrate problems created by trying to move ecosystems contrary to where they naturally would go. In all decisions, align your strategies with what would occur without your intervention. Think of ecological restoration as a process of assisting nature rather than controlling or guiding nature. While there are exceptions, they usually are costly and more likely to fail.

If you fight nature, prepare for a long and frustrating process. Early during restoration of Stone Prairie Farm, Steve and Susan sometimes found their plan leading in a direction different from what nature would have done. They quickly learned that it was nearly futile to swim upstream and so allowed nature to be their guide. For example, they tried to eradicate thousands of black cherry and box elder saplings that were introduced into Stone Prairie by wind, birds, raccoons, and other wildlife. As animals moved from habitat to habitat, the seeds were disseminated in their feces. Initially, Steve fought black cherry with brush-hog and herbicide, and even pulled up young trees. He and Susan eventually realized that where the trees grew, they shaded out the agronomic weeds. In these locations that were naturally savannas, black cherry provided a perfect opportunity to jump-start savanna restoration. After about eight years, seeds and plugs of native savanna plants, including other tree species, could be successfully introduced into black cherry groves. This turned what would have been a never-ending battle into a partnership with nature. Prescribed fire helped to thin the cherry and increase light availability to the developing savanna understory vegetation. Now a savanna of oaks, walnuts, hackberry, basswood, and other tree species, with a lush understory of native grasses, sedges, and wildflowers, grows beneath the scattering of trees.

In restoring abandoned agricultural fields throughout much of the East, native sassafras and sumac serve as *cover crops* for tree species that do not compete well with herbaceous species. Aspen serves much the same purpose when restoring hardwood forests in the upper Midwest. While these nurse crops will invade on their own given time, they can be encouraged to speed up the restoration of forest cover. This is described more fully in chapter 7.

## Recruiting Volunteers and Hiring Professionals

At Stone Prairie, Steve and Susan intended to do everything themselves, occasionally involving friends. However, they soon learned that they needed to hire help to accomplish some tasks. Without help, it would be nearly impossible to successfully restore all areas or progress on a scale that they thought most efficient. They first needed to hire someone with heavy equipment to remove the largest of the fallen box elder trees and regrade the stream bank. The trees were too large and dangerous to cut up with a chainsaw, and their old Allis-Chalmers tractor and newly purchased Kubota were not up to that and some of the other large tasks. They also hired and bartered to complete prescribed burning in a timely and safe way.

Hiring professionals to help design restoration plans, or technicians to implement restoration programs, depends entirely on your knowledge, experience, and what you and your family and friends can accomplish on your own. It also depends on the ecological scale of your projects, specialized equipment needs, and your timeline. For example, doing appropriate soil preparation and planting cover crops over the large acreages of former row-cropped lands on Stone Prairie Farm had to be done correctly to get prairie established. Steve and Susan could handle the seeding, but not the tillage and cover-cropping. They could stay on their schedule only by hiring a neighbor to do those tasks.

Volunteers are great for the occasional need, especially a planned event such as a prescribed fire. They love participating in prescribed burning, but hand pulling invasive plants can be an imposition on friendships. No amount of food or beer is sufficient recompense for the tiring, mind-numbing labor of pulling weeds, weekend after weekend. Handle that work yourself, or hire help.

There are many groups from which volunteers might be recruited, from Scouts needing merit badge projects to church groups wanting to restore a portion of God's creation, or even corporate executives who use a workday in natural areas restoration to increase esprit de corps. However, it pays to be careful when involving others.

When bringing in volunteers, design tasks that have a clear, accomplishable endpoint, ideally jobs that a few hours of work will likely get done. Do the same with hired help, but there the motivation for well-planned and organized tasks is focused on ensuring implementation efficiency to keep costs under control and projects on budget and on time.

Before hiring professionals, discuss your needs and the reputation of those being considered with natural resources agencies or landowners who are familiar with their work. Lists of professional businesses with the skills and experience you seek are usually available from The Nature Conservancy, or state and federal agencies who hire these same services for projects on public land. The prequalification process used by agencies to consider an organization may not provide enough confidence in the talents and attention to details you wish. Ask for references from previous clients and check out the reputation of professionals before you sign any contracts. For larger jobs, be sure to prepare and execute good contracts, making very clear the work to be done and the time it is to be completed, as well as payment and performance terms.

Many organizations, including The Nature Conservancy and local land trusts, will know of experienced volunteers and reliable professionals who may help with your restoration work. Steve and Susan found a wealth of talent among neighborhood farmers, many of whom were giving up farming and looking for alternative work. With good direction, they often developed creative, practical approaches that were less expensive then what had been planned.

With the restoration plan, management units, and budgets that you have prepared from the processes outlined in chapter 3, you should be able to determine how much and what kind of help you need, and do rough estimates of the costs.

## Restoration Success and Nature's Clock

The best-laid plans depend on nature. Restoration is no different from farming in this respect. You do all you can to increase your odds of success, but nature, particularly weather patterns, will greatly influence or entirely determine success. Without water, wetland restoration will fail. With too much precipitation, seed can erode from the carefully prepared field or rot in the soil. While these are the everyday risks, there are some important things that you can control.

### *Availability vs. Phenology*

Clock and calendar time are useful, but the real gatekeeper for timing of restoration projects is the growth cycle of plants. Phenology is the study of seasonal patterns of naturally occurring events such as animal migration or flowering of plants. For example, mowing to curtail weed-seed production needs to occur just ahead of seed set on target plants. Depending on the weather, the timing of growth and seed set may vary by several weeks from one year to another. Flexibility is an essential prerequisite for success.

Early in the restoration process, it is more critical to follow nature's clock and calendar and not miss the dates when management treatments are most effective. This is because you are trying to turn an ecosystem toward positive feedback loops, like trying to turn a large ship headed toward a rocky shoal. Initially, every inch, every second is important. Once on the correct course, you can relax a bit. If a weedy species gets a start for even one year, it may double or triple in population and present a far greater challenge the next year. Focus on jump-starting the restoration by being strategic, and timely, with your treatments.

### *Reading Phenology: What to Look For*

Pay attention to the plants and their life cycles. From flowering to seed set, most herbaceous plants require about three weeks. While beginning of flowering can vary from year to year, depending on rainfall and temperature, the three-week rule of thumb provides a pretty good projection for planning activities, whether it is management of weedy plants or collecting plant seeds. Most perennial plants have their lowest root reserves of stored energy just as they begin flowering, and this is a good time to mow or herbicide for optimum impact. This is especially the case for rhizomatous shrubs such as sumacs, prickly-ash, honeysuckle, and buckthorns, but holds for most perennial herbs as well. It is probably for this reason that an off-season, midsummer prescribed burn will often control undesired perennial vegetation better than spring or autumn burns. Getting the timing right and incorporated into your schedule of restoration treatments is a key to early success. It also is a key to maintenance, especially in the early years.

## *Knowing When It Is Too Late*

If you find that you are too late for a treatment, you have likely lost a year. For some species, this is definitely the case, while for others, you may be able to treat more aggressively and get by. For some species, back-to-back herbicide application, or fire in the autumn and again in the spring, can keep a project on track. The second-most effective period occurs when plants are actively mobilizing photosynthates for storage in roots. These are the products of photosynthesis—simple sugars that are linked together into larger starches that are stored by plants in tissues such as tubers or roots (think potato or carrot). The right herbicide will be mobilized in the plant with good results in late summer, while plants are still physiologically active. An herbicide treatment in the late growing season, followed by burning the following spring, is often quite effective. This back-to-back combination is helpful where your schedule does not align well with the plants' phenologies.

## *Throwing Everything at the Land?*

"Shock and awe" is a military approach that does not work with nature. Nature usually has more resiliency and survival strategies than can be overcome by brute force. Only by being strategic and good observers and learners can we ensure that our restorations will be successful. Being prompt and effective during start-up and keeping up with good monitoring and response to change are key to success.

## *Patience, Persistence, and Long-term Perspectives*

Some of the most exciting restoration responses may take many years to be revealed. Only after twenty years at Stone Prairie Farm did some of the first hand-tossed seeds, put in areas where soil was not well prepared, finally show up. Luckily, Steve and Susan did not start over in these instances, and they learned the importance of patience. Even now some very rare plants from the original seeding are appearing in their prairies and wetlands. This should remind us that nature is patient and persistent. To work with nature, one must be the same. Take a long-term perspective, and let go of immediate gratification.

Restoration projects most commonly fail when you lose sight of the lessons nature is trying to teach, especially patience and persistence. It is easy to let impatience push you into actions that are counter to natural processes because of the passion and excitement you have for restoring the land. There is nothing as exciting as a new wildlife species or new species of native plant that has miraculously shown up and now nests or grows on the restored site. Alas, people do not live long enough to experience enough of these revelations, but take comfort in knowing that you have started the system toward recovery, and those that follow will have these opportunities long after you are gone. Like life itself, ecological restoration is a journey, not a destination.

## Commonly Employed Techniques

A large compendium of techniques comprises the tool kit needed in ecological restoration. Most techniques aim to change the physical, hydrological, chemical, and biological processes on the land over time, in predictable ways.

### *Hydrology*

Restoration frequently requires restoration of hydrology, the way in which water enters and leaves the landscape. Scale can range from backyards to entire watersheds. Techniques range from disabling tiles to removal of levees or dams. Natural meanders may need to be restored in channelized streams, and grade controls may be required. The following are some common techniques used to correct hydrology.

**Grade control.** In efforts to dewater landscapes, the gradients of waterways and streams have often been increased. More rapid movement of water increases down-cutting of the channel. Increased runoff also increases down-cutting. Down-cutting results in elimination of meanders, and loss of related natural features. Grade controls are barriers placed in drainageways and channels to reduce gradients and encourage sedimentation (fig. 4.1). By retaining materials being moved in the channel, primarily during flood events, the channel begins to refill. Perhaps the best examples are beaver dams, grade-control structures responsible for millions of acres of wetlands. In forested streams, fallen trees are a natural grade control and help maintain

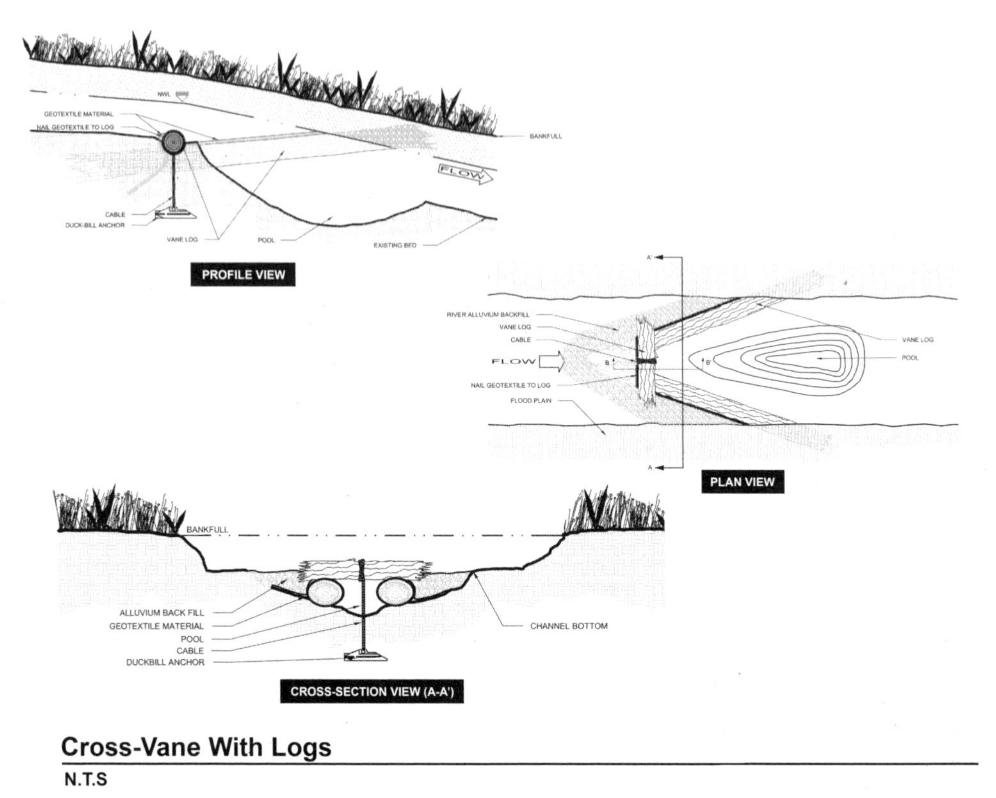


FIGURE 4.1 Grade-control structures can be simple log or rock features that are carefully installed in eroded drainageways to help rebuild the stability, restore the hydrology, and reestablish native vegetation communities. This image shows a cross-vane with logs that can be installed in perennial or intermittent drainageways to help capture sediments and rebuild stable grades.

a variety of pools that provide important habitat for aquatic organisms. Artificial grade controls can be used to prevent further down-cutting where excessive runoff from neighboring properties cannot be directly addressed. In ephemeral waterways, they can be as simple as log or rock structures, sometimes with earthen materials that are planted with heavy cover of strong rhizomatous grasses or shrubs to ensure their stability. Heavy stone from within the channel can be used, or even large trees can be felled across the channel and held in place by earthworks or stone. In urban or suburban waterways, concrete grade controls are often used.

**Interception and retention of water.** Water management begins before raindrops hit the ground. Vegetation intercepts precipitation, allowing it to more gently reach the soil. Litter and roots cover and bind the soil, thereby increasing infiltration and reducing runoff. In some areas, interception of moisture by vegetation is critical to the ecosystem. For example, in the Pacific Northwest (United States) and coastal environments south to northern California, fog interception by trees is an important source of moisture. Data from the midwestern United States show that intact prairies may have captured as much as one inch of rain with no measurable increase in soil moisture or runoff.[1]

Retention of water can also be increased in the landscape with small depression features that vary from hand-sized indentations to several-acre wetlands. Microdepressions are created naturally by activity of animals burying and digging in the soil, by invertebrates such as ants, and by wind acting on vegetation. Tip-ups caused by wind pushing over trees give rise to cradle-knoll microtopography in forests. Decaying stumps also form depressions that can be a meter or more deep. Fallen trees can not only form grade controls in streams, but create natural damming features on the land. Many seeps and springs are fed by water collected by the natural depressions that develop in healthy ecosystems.

Any activity that reduces vegetation cover or litter on the ground or organic matter and roots in the soil, or increases soil compaction results in less interception and greater runoff. Greater runoff almost always results in more erosion, setting up a vicious cycle. More erosion means less plant productivity and less infiltration. In semiarid landscapes, the reduction of infiltration can be quite serious. With reduced recharge, springs dry up and the water table will drop. When the restoration of native vegetation is implemented, springs and seeps are often restored.

Especially in agricultural areas where soil has been tilled, infiltration is often compromised. Tillage usually erases the microtopography created by vegetation and fauna. Practices such as cover crops, sod-waterways, contour tillage, and terracing are commonly used to reduce runoff and erosion, and increase water retention. During ecological restoration, physical modifications of the land can be retained or removed, depending on your budget and how far you want to go in restoring the natural landscape.

**Restoring underground hydrology.** Throughout much of the humid world, drainage systems have been installed to dewater land to increase plant productivity. While much of this alteration of the landscape is in conjunction with agriculture, some has been done to increase tree growth, such as in loblolly pine plantations in the southeastern United States. Dredging and straightening ditches and installing tiles removes water more quickly from the land. As soils are drained, oxygen becomes more available to roots, essential for good growth of most crops. Changing the oxygen level in soils greatly alters ecological processes, disfavoring the wetland species that would have occupied the poorly drained soil. The result of removing or backfilling drainage ditches and disabling agricultural tiles is a diffuse drainage network that retains water on the land, releasing it slowly.

Resaturating soils alters the chemical environment from oxidization to reduction, with concomitant changes in pH and nutrient availability. Nitrogen, for example, is soon lost as a gas, and phosphorous becomes less available. While these changes are bad for agricultural crops and many invasive species, they favor the native species that are adapted to wetlands.

In restoration, ditching and tiles should always be disabled unless they support drainage of adjacent properties. Disablement sometimes can be as easy as backfilling ditches into which tiles drain. Sometimes a backhoe can be used to dig up or break tiles. Subsoilers, heavy vertical blades pulled behind large tractors, also can be hired to break up tiles.

**Decreasing transpiration.** Invasive trees and shrubs result in greater transpirational water loss in many ecosystems, especially in semiarid and arid environments. Native species outside their original range and habitat may also become invasive. The reduction or removal of invasive plants can effectively increase water in dry ecosystems, both in the soil and in groundwater that feeds seeps and springs.

**Reconnecting hydrology systems.** In many locations, changes in water management have dewatered some areas while wetting others. Where flood-control dikes and levees are installed, the river is subjected to greater peaks and lows. Dams alter riparian wetlands because of disruption of the natural flood regimes. Virtually all major river systems in the United States have been altered by watershed disturbance as well as by structures intended to minimize flood damage. Ironically, areas that were intended to flood less frequently often suffer greater damage as a result of development behind the flood barrier and occasional breaching of dikes and levees. Restoration of streams and floodplains may require removal or disruption of dams, dikes, and levees, and restoration of watersheds and channels that were deepened or straightened.

Most rivers are connected to their riparian communities only during floods. Historically, floodplains received floodwater at intervals, depending on the climate and watershed. Flood events physically structured the river channel and floodplain. Floodplains relieve downstream flooding and improve water quality by removing at least some of the suspended sediments being carried by the stream. Dams, with or without dikes and levees, alter the flood regime. Removal of dams, dikes, and levees usually involves agreements between the state and landowners in the watershed. Legal challenges and approvals are generally a far greater hurdle than the removal of the structures. Keep in mind that permits and approved work plans are generally required for manipulating stream channels. (This is discussed further in chapter 9.)

**Restoring soil health.** Healthy soil is a living sponge. For every 1 percent increase in soil organic matter, each acre of land in the central United States will hold approximately sixty thousand gallons more water and remove six tons of atmospheric carbon dioxide that otherwise would contribute to climate change. Water is captured by humus and organic matter in the soil, from which plant roots can extract it. Excess is slowly released to percolate into groundwater. Healthy soil, a product of a healthy ecosystem that also intercepts precipitation as described earlier, significantly reduces surface water runoff and increases infiltration. Healthy soil encourages groundwater infiltration, restoration of groundwater levels, and provides base flows for springs and seeps, maintaining wetlands, streams, and lakes.

Rebuilding soil structure where tillage or grazing has damaged it can be done quickly in many ecosystems. Initially, quick-growing cover crops can be established on depleted agricultural fields. In some instances, these crops can be tilled into the soil as *green manure* to provide a quick increase in organic matter and to control weeds. Green manure is a term applied to short-lived, fast-growing vegetation, such as buckwheat, that can be grown and tilled into the soil. The same or different cover crops can be used to nurse establishment of native species. Natural vegetation, especially grassland and wetland vegetation, once established, will continue to contribute to soil development and protection. It has two benefits over cover crops, which are usually short-lived, often annual, species. Natural vegetation will accumulate mulch to further protect the soil, and equally or more important, it will provide critical habitat for myriad invertebrates and other animals, not to mention microbes, that are necessary to build and maintain healthy soils. Studies have shown that compaction of soil by heavy grazing can persist for many decades after removal of livestock but can be restored to reasonable health in two or three years with proper management.

**Adjustments for landform.** Landform, the topography of the landscape, can make a surprising difference in available moisture, and this can greatly affect planting dates and techniques required for successful establishment of vegetation. Effects increase with latitude. On southern aspects and steep slopes, available moisture during the growing season can decline by the equivalent of several inches of precipitation. Adjustments often are needed in such exposed settings, and include planting earlier to capture spring rain or snowmelt, mulching, or use of cover crops. Microindentations in the surface, often required in desert planting (chapter 10), may be necessary on these excessively exposed sites.

**Intercepting off-site water.** In agricultural and urban environments, much of the runoff from adjacent farms or impervious surfaces such as parking lots can be accommodated nicely in restored landscapes by designing catchments with native vegetation. Effective strategies to capture, hold, and cleanse water also can facilitate infiltration and evaporation, thereby reducing total surface runoff volumes. Examples include rain gardens and plantings of deep-rooted native species through which water can be routed. Size and number of catchments can be adjusted depending on the volume of water requiring interception. These techniques can be incorporated into any green space, including road rights-of-way, medians, lawns, parks, and vacant areas, which can be landscaped to favor birds, butterflies, or amphibians (see chapter 11).

## *Chemical Challenges*

Chemical problems are generally one of two types: contamination by waste materials or alteration of pH and nutrients beyond the range of natural variability found in healthy ecosystems. The latter is often associated with change in soil drainage but can also result from acid deposition from coal combustion, fertilizer spills, or runoff from feedlots. Adjustments to soil or water chemistry may be relatively simple. For example, when soils are rehydrated by disabling tiles or ditches, a reducing environment is established and this typically lowers pH. On the other hand, prescribed fire usually will increase pH by ashing plant residues and releasing calcium and magnesium. Use of agricultural lime also is relatively easy and inexpensive. On the other hand, mitigation of toxic chemical contamination can be costly and difficult. To assess the problem, samples of soil or water, and often plant and animal tissues, must be collected and analyzed according to a rigorous set of protocols. Even then, interpretation requires reference to background levels, so the assessment is typically done by a commercial laboratory.

Sources of contamination frequently arise from adjacent property, particularly highways, parking lots, or agricultural or industrial operations. Highway salt, for example, is a particular insult to freshwater streams. High levels of nitrogen and phosphorous from adjoining property often can be addressed by creating or restoring wetlands that act as biofilters. Artificial wetlands can be created in shallow scrapes with small one- to two-foot high contoured dams on intermittent drainageways vegetated with wetland species. If nutrient loading is excessive, accumulated nutrient-enriched sediments may periodically be spread on agricultural fields during dry periods, and the depressions replanted.

Bioremediation, the use of rapidly growing plants to absorb contaminants from the soil or water is being used more commonly. If the contaminant is fertilizer, the plants can be harvested and used for fuel, fiber, or mulch, or composted and used as a soil amendment. Use of plants to absorb toxic substances from the soil is also increasing. Plants can either stabilize the contaminant on-site, or be harvested and treated appropriately, even as hazardous wastes, if necessary. Plants stabilize the site and increase organic matter, which increases the capacity of the soil to tie up contaminants. It is often much less expensive to use plants to remove or immobilize contaminants than other remedies.

Biofilters are one of the easiest and cheapest ways to clean up runoff water. Biofilters also increase water retention. Biofilters range from vegetated waterways to strips of forests along streams. Water emerging from the biofilter is cleaner and cooler.

In most cases, removing the source of the contamination is all that is required to restore contaminated streams, as a stream is an open system and will flush itself, given the opportunity. Slow-moving streams, ponds, and wetlands, however, require much longer to flush. Sediments in pools can persist for decades. Because of the risk to downstream communities and the cost, it often is prudent to be patient rather than using mechanical intervention to remove contaminated sediments. Sometimes the easiest fix is to aerate the water, allowing the increased oxygen to stimulate decomposition.

Reducing or eliminating the source of chemical contamination is often essential. Dilution may be the easiest solution if the contaminant is not toxic or does not bioaccumulate. Fertilizer spills are routinely handled by removing contaminated soil or water and spreading it over cropland at very low concentrations. On one project, excessive phosphorous originated from a five-acre area formerly used to dump manure from a turkey farm. The contaminated soil was spread lightly over hundreds of acres of farm fields, which were then converted to prairies, savannas, and wetlands. Phosphorus was effectively tied up in the developing tissue and roots of the restored ecosystems, then released slowly in the nutrient cycle.

It may be useful to have some locations in your project that are effectively "throw-away" restoration areas. Biofilters used to intercept nutrient-rich runoff from the neighboring farms may essentially be sacrificial areas. Invasive species may be nearly impossible to keep out. Sometimes contaminated areas can simply be covered with soil and planted to native species, with no follow-up required. In one project, we successfully capped an eroded half-acre feedlot with topsoil and planted native species. In a paved feedlot at another project, the concrete was fractured with dynamite to allow infiltration, then the area was covered and planted. Where it is impossible to ensure containment, we have installed a combined surface and subsurface treatment train that alternates between aerobic (unsaturated soil) and anaerobic (wetlands) treatment strategies that accomplishes a very high level of chemical removal. Chemicals that are not broken down in the oxidizing environment are likely to precipitate or be off-gassed in the anaerobic reducing environment (for example, nitrogen gas is released under anaerobic conditions).

Many restoration projects have some debris cleanup, removal, or composting and burial needs. These can usually be accommodated on-site, but be mindful of chemical residues that are often associated with debris. Burial might take unsightly debris out of view but contaminate groundwater. Likewise, burning can release contaminants into the air. When in doubt, check with professionals about toxic substances, especially when dealing with old building debris or equipment.

## *Biological Techniques*

Reintroduction of native species is nearly always a component of ecological restoration. This involves both direct and indirect management strategies. Many indirect techniques are aimed at reducing invasive species that compete with or prevent the establishment of natives. Sometimes other species are used to combat the aggressors, as in the case of cover-cropping or biological control agents. Examples vary from herbivores, including insects, to microbes and plant diseases that can help reduce or eliminate an invasive plant species. Cattle, for example, have been shown to maintain oak savannas in the midwestern United States. Also, recall the indirect technique at Stone Prairie Farm, where invading black cherry saplings were allowed to shade out agronomic weeds. In another project, Alan made brush piles in a forest where red maple was becoming overly abundant in the absence of fire. Brush piles allowed a larger cottontail rabbit population, the rabbits girdled the maples during the winter, and attracted more owls and foxes, achieving yet another dimension of diversity.

Some biological techniques involve controlling wildlife that is preventing the growth and development of healthy plant communities. White-tailed deer, cottontail rabbits, and European carp, among other species, are typically managed through exclusionary fencing or barriers. Given the opportunity to become well established, native vegetation often will flourish without continued exclusion of herbivores, although there are exceptions. Many forest species have been prevented from regenerating in forests of the eastern United States because of unnaturally high deer populations.[2] Sometimes the introduction of seeds or plants is necessary to jump-start plant communities back to health, especially where no adjacent community contains a diversity of native plants whose propagules can spread into the restoration area. This may also be necessary when wildlife cannot be counted on to disperse the plant propagules in time to stabilize the site against invasion by weedy plant species.

**Invasive species.** The most common biological problem in restoration is control or eradication of invasive species. Invasives may be either plant or animal, nonnative or native, species that are adapted to disturbed or dysfunctional ecosystems. Techniques to control invasives can be as laborious as digging or hand pulling, where the numbers are few and the area small, or as extensive as introduction of an herbivore or disease, or broadcast spraying of herbicides to kill all plants before attempt is made to establish the desired plant communities. The latter is most common on agricultural land where weedy propagules have accumulated, or where persistent weedy perennial cover has been allowed to develop.

Small areas of invasives can be covered by weighted-down heavy black plastic sheeting for one or more growing seasons, followed by seeding or planting with native species. In the Midwest, where glyphosate-resistant soybeans are routinely grown, agricultural fields infested with weedy vegetation can be planted with this crop, followed by glyphosate applications to eliminate weedy vegetation. After soybeans are harvested, native species can be seeded into the soybean stubble in the fall or following spring. This is especially useful for establishing prairie.

Another useful technique involves overwhelming invasive plants with "smother crops." At Stone Prairie, in an area with persistent perennial stinging nettles and reed canary grass, soil was tilled to encourage giant ragweed for several years. The ragweed formed dense growths that towered over nettles and canary grass and suppressed them. The area was then burned and native grasses and sedges were introduced. Cover crops, including trees, grasses, or forbs, can serve the same purpose by either occupying newly exposed soil before invasives can get established or shading invasives that require full sunlight. Cover crops are also used to secure newly prepared sites against erosion and to protect from freezing and thawing and rapid drying, thereby favoring establishment of native species that are slower to get started.

**Herbicides.** For many people, herbicides are a double-edged sword. On one hand, modern herbicides are effective and, properly handled, of low risk to humans or wildlife. On the other hand, they represent chemical contaminants, often with unknown side effects, in contrast to the natural processes owners wish to restore. Fortunately, herbicides have been greatly improved over the 2,4-D and 2,4,5-T nonspecific herbicides widely used in past decades. Glyphosate, in particular, has a short half-life in the soil and very low toxicity. Given proper care in application and selection of chemicals with very low persistence or toxicity, the decision on use is primarily philosophical for many landowners. Need for herbicides generally decreases as healthy ecosystems are restored. The type of application is important. For example, wicking reduces the herbicide quantity (costs) and allows finer discrimination on which plants get treated. Selecting the best herbicide and properly applying it can greatly increase the rate and overall success of ecological restoration where control of invasive or competing vegetation is required. Note that many herbicides are restricted or require a license for applicators, and those often would not be your first choice anyway. Nevertheless, all herbicides should be considered potentially hazardous, and handled according to directions on the label.

## *Site Preparation*

The aim of site preparation is to favor establishment of native plants and diversity that at least approaches natural assemblages. Often, restoring native plants first requires reduction of weedy plants. Control of nonnative and invasive vegetation is necessary during all phases of establishment and maintenance. Herbicide or prescribed burning, as described later, with or without soil preparation, are the primary methods of controlling competing vegetation prior to planting. Mowing and brushing are more commonly associated with early establishment and maintenance. Here we discuss only soil preparation.

If the soil on a site has not been greatly damaged by erosion or chemical contamination, it can largely recover in a few years under proper management. Otherwise, healing of the soil can take decades. Soil development in most cases takes centuries, so it should be no surprise that repairing damaged soils is seldom quick. Although the process of soil improvement can be speeded up, there really are no shortcuts in the establishment of a healthy terrestrial ecosystem. Soil health and ecosystem health are two perspectives of the same objective.

Do not assume that as long as plants have water and good soil contact they will thrive. Take time to ameliorate soil that is too acidic or too alkaline as much as possible before investing heavily in seeding and planting. Also, investigate thoroughly, as described in previous chapters, to make sure the right species are being sown in the right locations. A lot of valuable time and seeds can be wasted by planting them in the wrong soil. At Stone Prairie, bags of hard-earned native seeds were initially tossed on soils that were deficient in organic matter and compacted because of years of intensive farming. Some species did not show up for a decade or more, whereas in healthy soils those species would have established within two or three years. When native species are slow to become established, weeds are given a competitive advantage.

Weedy species have the advantage in eroded or compacted soil. Disking once or twice during the growing season to encourage germination of weeds, thereby reducing the weed-seed pool before seeding native species, will work if soils are reasonably healthy. If soils are compacted or low in organic matter compared to what would be found in natural remnants of the same community type, consider green-manure cropping for a few years, or application of organic materials. Nonweedy legumes such as ladino clover can also be used in some regions. Clovers have an added benefit of adding nitrogen, although that is not usually needed or desirable for native species, and their deep roots help to break up compacted subsoil. A range of other plants are often worth considering to quickly help rebuild soils in temperate regions:

sorghum, winter rye, barley, and oats. Use deep-rooted species, such as canola, when soils are compacted.

When soils are badly compacted, as is often the case when clay-rich soil has been grazed heavily for long periods or regularly plowed, deep chisel-plowing may be necessary. Subsoiling is sometimes needed to break up deep impervious layers. A subsoiler is a heavy, curved point that extends two or more feet into the ground, pulled behind a tractor. Subsoiling also can penetrate deep enough to break up tiles if pulled with a heavy tractor or bulldozer, accomplishing multiple tasks with one effort. The local NRCS or agriculture extension office can generally direct you to contractors who can do this work.

## *Seeding and Planting*

There are many techniques for introducing plant species to a site. When it seems unlikely that native species will develop from the seedbank, seeding or planting is required. Planting is much more labor intensive and expensive, so seeding is preferred for large acreage and for species that will form the primary matrix of new vegetation. Establishing a diverse ecosystem is a long-term task. You might want to conceptualize the process as similar to how an artist creates a painting. Usually a base coat of paint is applied. This is equivalent to sowing a cover crop. While this can be sown at the same time as the other layers of plant seeds, many perennial species will not establish easily, if at all, until a plant matrix of native species is established. The cover crop can be planted first, to stabilize soils, ameliorate soil moisture and temperature conditions, and stabilize soil chemistry. Additional seeding layers can be applied later. We recommend adding seeds of early-successional and short-lived perennials initially. The more conservative species, slow-growing perennials, can be seeded or plugged into the restoration once native vegetation cover is established. Depending on the species to be added, mixes of complementary associates can be added together. For example, where transitional sedges should dominate, above the normal water level but within seasonally saturated wetland soils, one might sow mountain mint, various sunflowers, and rudbeckias, much as an artist would choose different pigments in appropriate sequences.

Consider the pollination requirements of the species being added to a community. If the plants have obligate insect pollinator requirements, attempt to seed or plug new stock to establish critical masses that will attract pollinators. It is delightful to watch sphingid hawk moths (family: Sphingidae) or solitary native bees pollinating orchids, shooting-stars, and prairie and Turk's-cap lilies, but these insects are not effective pollinators where host plants are isolated. They do best when a critical mass of nectar-producing plants is established. When you witness frequent pollination of forbs by less-common pollinators, you can be confident that you have made great progress toward reestablishing a healthy ecosystem.

The timing of seeding can be critical to success, but varies by ecosystem. Obviously, seeding on frozen ground or during a dry season will not yield prompt results. However, dispersing dormant seed during either of these times can ensure that the seeds will be ready when proper conditions for germination occur. In the Midwest (United States), clay soils can be very wet in early spring and then become hard and dry by midsummer. It is often difficult to get into fields with equipment during the wet spring period, and the window of opportunity narrows rapidly as summer approaches. Therefore, it is less risky if seeds are sown in late autumn of the preceding year, as a dormant planting. This involves drilling or broadcasting seed after cold weather arrives so that the seed does not germinate until the following spring, after the soil becomes sufficiently warm and moist. We also have seeded directly onto snow, by broadcasting the seed, then running over the crusted snow with a drag or disk pulled by a tractor to make sure the seed does not blow away or is not consumed by birds. If doing this, be sure to till the soil well before snow arrives. The seeds will be planted in the spring by freezing and thawing as the snow melts.

**Seeding.** Seed is typically sown using conventional or specialty drills, or by hand-broadcasting. Drills can meter seeds to ensure relatively even dispersal. Hand-broadcasting is difficult over large acreage and often results in more spotty distribution. A mechanical cyclone broadcast seeder can be a good alternative. They can be obtained in various sizes, from hand-held seeders that hold a few pounds of seed to larger varieties that can be attached to an all-terrain vehicle (ATV) or, with a three-point hitch to a farm tractor. Specialty drills also can be rented in many areas and used in sod that has been killed by herbicide late in the preceding growing season. This avoids disturbing the soil and stimulating germination of weed seeds, and it also keeps soil covered to protect it from drying and wind erosion. Check with your county conservationist to get their recommendations for finding a drill.

With either drills or cyclone broadcasters, multiple applications are usually not necessary to deliver seeds of different species, since these pieces of equipment can handle seeds that vary greatly in size and ease of spreading. Be aware that many native plant seeds are light and fluffy, with awns or hairs that prevent them from flowing through conventional agricultural drills. If you are broadcasting, they will not spread far, except on the downwind side of your track. This includes most native warm-season grasses. Some commercial sources offer debearded seed that has had the awns and hair removed, thereby making it easier to sow. Some specialty drills are designed to handle low-density, fluffy native seeds. You may need to contract with a commercial operator for seeding large acreages.

The key to drilling or broadcast seeding is to ensure that the seed and soil come in good contact, meaning literally that the seed lands on or works its way into the upper surface of the soil, or at least lies in direct contact with the soil surface. Many seeds will germinate only at the soil surface, as they require light to trigger germination. Others require burial where prolonged moisture can be imbibed. Properly adjusted and used, drills can meter seeds onto the surface, for those that require light, or insert seed through at a range of depths. Generally, it is better to sow seed that requires light for germination directly onto the soil. When broadcasting on the surface, follow with a drag, harrow, or cultipacker to increase contact between seeds and the soil. Disking is not advised because too many seeds that can germinate only near the surface will be buried too deeply.

Broadcast seeding is sometimes better than drilling when you are sowing a variety of species. After preparing the soil by disking or tilling, broadcast those species that require burial, and follow with very shallow disking or harrowing. Then broadcast the species that need to remain at the surface, and follow with a cultipacker or drag to firm the seedbed. If you are sowing in late autumn before the ground freezes, freezing and thawing during the spring thaw will settle the soil and ensure that seeds are in close contact with the soil before germination begins, in which case you can forgo the dragging or cultipacking.

In areas where heavy cover of perennial grasses or alfalfa is present, herbicide is usually necessary before seeding. No-till drilling can follow in the autumn or early spring, if a specialty drill is available. Sometimes, no-till drilling can be done directly into the mowed sod or perennial vegetation, with follow-up herbicide before the native species germinate. If planting into fallow fields, high mowing can be done to release the germinating plants. During droughty years, the no-till method is especially attractive because it conserves soil moisture. Killing perennial forage vegetation often releases annual weeds such as foxtail grasses. The foxtail can be used as a free annual cover crop into which native seeds are drilled or broadcast. This is the same as if direct seeding were done into annual cover-crop grasses planted for that purpose.

**Seed sources** for the United States and Canada are now readily available with an online search. We encourage using seeds from remnant natural areas near the restoration site when possible or working with a reliable native plant nursery that will assure you that seeds and plant stocks are of reasonably local genetic stock. Because of genetic variation, the more local the seed, the more likely the variety is suited to local climate and soils. Check to be sure, however, that seed collection is permitted, and secure permission from landowners or managers. Hand collecting is an excellent opportunity to involve volunteers or extended family, and it provides a perfect opportunity to see uncommon species that might add diversity to the restoration project. Mechanical harvesting equipment such as combines and flail-vacs can be used where there are large patches of native plants, and equipment can be safely operated.[3]

**Planting.** If you do not grow plants from the seeds you have collected, you can buy native plants from commercial sources. You can also collect and grow your own plants, or consign a native plant nursery to grow plants for you from seeds you have collected. It is common to use seeding to establish the primary vegetative matrix, then use plugs or plants to introduce less-common species to augment diversity. More mature plants provide quicker results, but cost more and require more time to plant.

Although less so than with seeding, timing is important. If you plant too late in the growing season, you may need to water new plants to carry them through hot, dry periods. If you plant too early in northern areas, particularly in heavy clay soils that hold water, frost heaving can pull young plants out of the ground before they are well established and rooted. Frost heaving is caused when ice forms at the surface around stems. As freezing continues, the plant is lifted from the soil. When the ice crystals melt, the stem is released, but the thawing soil will have settled. Three or four freezing nights followed by daytime thawing can pull even young trees nearly free of the soil. We have witnessed thousands of plugs lying nearly root bare on the surface, exposed to sun and drying after being frost heaved. To avoid this problem, plant a bit later, plant into a cover crop, or mulch the plantings.

When installing plugs or container-grown plants started in a rooting medium, planting involves digging a hole, popping the plant from the container, placing it into the hole, and pressing the soil around the plant to fill the hole. Bare root stock, which is less expensive, usually can be planted by cutting a slice into the ground with a dibble or spade, into which the plant is installed with roots carefully spread, followed by tamping the slit closed around the plant.

Trees and shrubs also can be purchased as containerized, bare root, or balled. Balled stock is usually wrapped with burlap or other breathable fabric that holds soil and moisture around the roots. At the time of planting, remove or at least cut and peel back the top of the material so it doesn't constrict stem growth. For all stock, spread the roots into the hole you have prepared. Avoid mixing topsoil and subsoil when you are backfilling soil around the newly planted materials. As much as possible, keep topsoil close around the newly planted stock. It holds water and nutrients better.

Bigger plants require bigger holes, somewhat larger than the root ball or container. Larger stock is more expensive, and survival is often poorer. It also requires more digging, unless you have access to small augers or backhoes. A smaller tree planted correctly in the right location will usually outgrow a larger tree improperly planted. This applies to nearly all trees, but especially hardwoods such as oaks, walnuts, maples, and hickory. Consider one- or two-year-old seedlings or cuttings rather than older stock, unless there is some special requirement for the immediate presence of larger trees.

Most trees, shrubs, and vines also can be planted as seed. Excellent forest restoration projects have been established with direct seeding into lightly disked soils with a nonaggressive cover crop. In some parts of the United States, large-scale direct seeding has been accomplished by mixing seeds of different tree species with old sawdust, chipped brush, leaves, or even ground corncobs and broadcasting with a manure spreader. In some cases, the resulting reforestation has surpassed the results of meticulous hand or machine plantings. Some handbooks describing direct seeding are listed in the endnotes.

Continuing the artistic analogy, the final details of installing a restoration are added over time as a part of the diversification strategy, much as the artist will add final details that enrich the painting. Whereas direct seeding is ordinarily done initially, or in the first year or two, adding planting stock can be done later. In every project, there are opportunities to increase genetic and species diversity through proper timing and sequencing of seeding and planting. At Stone Prairie Farm, Steve and Susan continue to enrich the genetic and species diversity each year by adding plugs of the more rare species they grow from seed.

## *Control of Woody Vegetation*

It is often necessary to remove or reduce native and nonnative woody plants. Hand sawing is physically demanding and can be dangerous. Many woody plants will sprout after stems have been cut, often aggravating the problem of control. During site preparation, a root rake on a dozer can be hired, but that is not an alternative on small areas or after native species have been introduced. Root rakes, commonly used to prepare sites for reforestation or preparation for construction, are comparable to reworking the landscape with earth-moving equipment. Heavy prongs on a dozer literally rip through the soil to dislodge root masses that are then pushed and piled to be burned or hauled off. Brush-hogs require heavy tractors, scarcely disturb the soil, but leave all the stems and roots to resprout.

**Brushing.** Brushing must be carefully planned to match the methods to the species. Repeated treatments usually can be avoided by knowing the response of the species you want to reduce or remove. Most broad-leafed trees and shrubs will resprout after they are cut. If not treated, nearly every stem will develop multiple sprouts from the stump or from roots. Species such as quaking aspen, black locust, or ailanthus will give rise to a new forest of saplings covering a much larger area than the original area you treated. Application of a woody herbicide, such as triclopry or piclorum, needs to be applied to broad-leafed stumps soon after cutting. Little resprouting will occur after cutting most conifer species.

Where small saplings or brush are growing in old farm fields, nothing beats the heavy-duty brush-hog mower that most farmers have in their machine sheds. Even small trees can be reduced to chips with use of a Hydro-Ax, an industrial brush-hog mounted on a crawler or skidder. While expensive to hire, the investment may be cheaper than the laborious task of clearing the land by hand, but be prepared for the time and expense of applying herbicide to the stumps. If your objective is to replace the woody vegetation with herbaceous species, there is no alternative to herbicide initially. Once fire-tolerant herbaceous vegetation is established, regularly prescribed fire can generally prevent reestablishment of woody plants.

If you need to kill the woody plants, girdling will usually be sufficient with or without proper follow-up herbicide application around the trunk, depending on the species. Many trees will die if only girdled, a technique done with an ax or chainsaw. The idea is to completely disrupt a circle of phloem and cambium around the tree, preventing sugars from the leaves being transported to the roots. Some species can be killed when the base of the stem is covered with an appropriate herbicide; this will also prevent these species from resprouting. The basal bark application can be done initially with application to targeted stems using a hand sprayer, wicks, or sponge applications. Herbicide is injected into the stem with a special tool called an injector, or applied in a frill cut made with a small ax or machete. After the stem dies, often a year or more later, depending on the size and species, mowing or even prescribed burning can be used to clear the site. Alternatively, you can leave the standing dead woody plants, which will eventually fall or become consumed through prescribed burning and decomposition. Nearly all extension or NRCS offices will have flyers describing proper techniques and recommendations for herbicides.

Larger tree thinning or removal projects may be equivalent to logging operations. These can be achieved, sometimes through a timber sale. For example, in oak savanna restoration, it is not uncommon for half or more of the basal area of trees to be removed, with what foresters call a *low thinning*, or thinning from below. This means the smaller trees, usually more shade-tolerant species that have been established in the long-term absence of fire, are removed. If there is sufficient volume and a local market, a buyer can be contracted to do this work. Sometimes, large amounts of merchantable trees such as black locust need to be removed. This species is useful for fence posts, firewood, cross-ties, and pallets, among other things. If there are no merchantable trees, woody vegetation can be cut, piled, and burned, or chipped and scattered. Care is needed, however, because burning can sterilize the soil beneath the brush piles, opening avenues for invasives to become established. If chips are too thick, they can smother the herbaceous species you are trying to release. Consult with your county forester to see if you have enough volume to justify a timber sale. If you are unsure of which approach to take, try a technique on a small-scale test plot, and monitor results before scaling up.

The ultimate goal of brushing is increasing light availability (quantity and quality) to the ground so that ground-layer plants can grow. Avoid proceeding too timidly. It is possible to open the canopy sufficiently to stimulate germination of ground-layer species but not provide them enough light to survive. That can reduce the seedbank. If you plan to reduce woody vegetation to encourage ground-layer species, do it boldly in one operation.

## *Use of Fire*

Few restoration and management techniques are more widely applicable and have greater efficacy than prescribed fire. Fire, however, requires special training to manage safely and effectively. Steve planned to use fire to restore and manage the eighty-acre prairie, savanna, and wetland complex at Stone Prairie Farm. With their first prescribed burn, he and Susan learned that neither they nor the land were ready for fire, and a neighbor's barn was threatened as a result.

Other than in the humid tropics and rainforests, most ecosystems have evolved with repeated fire. Some were intentionally set by Native Americans, who recognized benefits of fire. They used fire to drive game, restore good grazing and browsing conditions to favor game species, and keep a site open to favor wind and discourage mosquitoes and other biting insects. They probably also used fire to maintain more open conditions in forests around villages for better visibility. Fires also were ignited by lightning. The fires that burned over the larger landscape generally occurred during the more extreme conditions. Ecosystems that were routinely subjected to hot, dry, windy weather were more frequently affected. Species in these communities became better adapted to fire. Indeed, fire is essential to the health of many types of ecosystems, from grasslands to savannas, to conifer-dominated forests, and even most broad-leafed forests, especially where oaks are dominant.

Fire triggers a new successional sequence and releases accumulated nutrients, stimulates germination of many species, reduces germination-inhibiting chemicals, frees up nutrients, exposes the soil to sunlight and increased temperatures, raises the pH, reduces woody vegetation, creates snags and coarse woody debris, and selectively disfavors species less adapted to fire. While there are no perfect substitutes for fire, some of the basic functions offered by burning can be achieved by mowing prairies, or by mechanical cutting or thinning of savanna and forested systems. We are not encouraging you to race out and start burning your property, or worse yet, someone else's if your fire gets out of control. However, the use of prescribed fire is often critical to successful restoration. It must be conducted carefully and strategically, and timed appropriately to achieve desired ecological results.

When using fire, it is important to recall the principle of ecosystem-effective scale. As much as possible, fire should be applied in the way that fire burned historically, allowing it to sweep across the landscape, burning where it will, and skipping what it won't. Too often, managers take a fire prescription too literally. If the prescription calls for fire, they assume every square foot should be burned, including all organic matter above the soil. To the contrary, prescribed fire should draw out the diversity of the land, not reduce everything to ash. We recommend starting burns in the early evening and burning into the night. This use of fire creates heterogeneity, or patch diversity. By burning when the relative humidity is increasing and wind speed decreasing, as is typical at nightfall, over the course of the burn some areas will be completely scorched while others do not burn at all. Fire variability translates into vegetation diversity. Unburned patches also provide important refugia for insects and other fire-sensitive animals. Also, after a burn, there is an opportunity to introduce additional species into the burned patches, increasing the biodiversity of the community.

Before doing a prescribed burn, make sure you have the right attitude, equipment, training, and help. You also need good information about pending changes in weather conditions and how fire will react to the changes. By attitude, we mean humility, ensuring the margin of prudence for safety. There are two classes of people who do prescribed burning: those who have had a fire escape, and those who will. We all want to remain in the second category for as long as possible. Fire demands respect and care.[4]

There are many sources for learning how to use prescribed fire safely. Organizations such as Applied Ecological Services Inc. or The Nature Conservancy (TNC) frequently offer training opportunities with both classroom and field instruction. A good starting point is the federal *Interagency Prescribed Fire Manual*, available online at http://www.fs.fed.us/fire/fireuse/rxfire/rxfireguide.pdf. Another excellent source is TNC's *Fire Management Manual,* also online at http://www.tncfire manual.org/guidelines.htm. You may be able to volunteer your assistance with TNC or other land trusts that routinely do prescribed burning in your area. *Experience is essential, so if you do not have any, get it or hire it.* Nearly everywhere in the United States and Canada there are experienced people willing to lend a hand. Be wary of thrill seekers, and choose your help wisely. Ideally, you will want assistance from someone with training that comes only from working on a fire line.

We have learned that altering the timing of fire within a growing season can be beneficial. Conducting spring, summer and late-summer, or fall burns rather then all spring burns will encourage diversification. We primarily use spring and occasional autumn burns, and rarely midsummer or off-season burns. Timing is always important. Spring burns done too early stimulate undesirable, cool-season grasses like smooth brome and Kentucky bluegrass. Done too late, fire can set back the growth of native species for the year of the burn. Midsummer burning is especially damaging to shrubs that are fully leafed out, a desirable effect if you are trying to open a woodland or savanna. Growing-season burns have been successful in controlling resprouting of woody and semiwoody plants you are trying to reduce or remove. Bear in mind that fire work is hot and physically demanding, so burning during cooler seasons has advantages for people on the fire lines.

Deciding how much landscape to burn at any one time is an important decision. We recommend that you always aim to burn at as large a scale as possible, ultimately limiting the fire to what you can control. Burning hotter is always better than burning too lightly, as long as you can control the boundaries of the fire. Prescribed fires nearly always are done when conditions are not extreme: low wind, high humidity, and moderate temperatures. This is for safety. As a result, prescribed fires across a diverse landscape will burn with patches that vary from no fire to very hot fire. At Stone Prairie Farm, burning mostly at night resulted in about 40 percent of the land remaining unburned after a normal spring fire. In heavy fuel areas, burning is more complete. We recommend burning safely, when fires are more easily controlled, and burning more frequently. However, lighter fire does not always get the job done, and either a hotter fire or combinations of fire and mechanical removal of woody vegetation is necessary.[5]

## A Management Program

As the ecosystem recovers, the project will transform into a long-term management program. A long-term management plan is needed to protect the investments made in the recovery of the ecosystem.

Ecological restoration is usually applied to ecosystems that have been degraded or damaged and require assistance for recovery. This early phase of restoration is often called *remedial restoration*. Even healthy ecosystems, however, require various degrees of maintenance, if for no other reason than to control invasive species. This is often called management or *long-term maintenance*. In ecological restoration, the management phase begins after the initial ecosystem recovery goals are achieved, or at least are on a trajectory toward success. Management has different goals and objectives. Monitoring informs us of when we have achieved the remedial goals, and how much effort and expense will be needed in the forthcoming years to address the long-term management goals. Management goals are typically focused on maintenance of a set of conditions that have resulted through the successful implementation of the remedial phase.

**Remedial phase.** This phase of restoration is when major efforts are undertaken to restore structure and biological diversity and begin restoring ecological functions. This phase includes such tasks as reducing nonnative and undesirable native species, restoring hydrology, and establishing native plant communities and species. It may include prescribed burning and usually involves removing or mitigating physical or chemical stressors.

**Management, or long-term maintenance.** After achieving remedial goals, the restoration process shifts to a lower-cost, reduced-intervention program. This phase includes tasks such as spot herbicide treatments, remedial planting, and prescribed burning. This provides a close-hand opportunity for long-lasting personal involvement by local residents, friends, and your family to stay involved in restoration of the land.

Once the ecosystem has reached the point where it has regained functionality, including the ability to respond to normal perturbations, it is still necessary to protect the ecosystem from outside threats. The level of effort and resources required is generally a fraction of what was needed in the first few years, during the remedial restoration phase. Management activities should continue to be planned for each designated management unit. Appropriate management tasks will be guided by regular monitoring within each unit, although the frequency of monitoring may be less.

A scheduling table, which you prepared before beginning restoration, and that should be updated each year to guide the annual plan of work, will continue to guide work over the long term. Because many management tasks may be repeated periodically for different ecosystems, it may also be useful to have a simple table that summarizes the frequency with which you intend to repeat specific maintenance tasks (table 4.1). This table will indicate your intentions, but will be subject to change

**TABLE 4.1**
*Treatment schedule, in years*

| Plant community | Prescribed burning | Spot herbicide treatment | Remedial seeding and planting | Monitoring |
|---|---|---|---|---|
| California live oak savanna | 2–3 | 2–3 | 3–5 | 1 |
| Dry prairie | 3–5 | 2–3 | 3–5 | 1 |
| California coastal grassland | 2–3 | 2–3 | 3–5 | 1 |
| Riparian wetland | 3–4 | 1–2 | 3–5 | 1 |
| Emergent wetland | 3–4 | 1–2 | 3–5 | 1 |

based on monitoring results. For example, you may plan prescribed burns on two- to three-year, three- to four-year, or five- to seven-year cycles, depending on the ecosystem. Note that monitoring should be conducted regularly, whereas management tasks are done as required. During long-term management, some monitoring measures or observations may need to be taken only every few years. Be careful, however, that you do not let monitoring slide. Maintenance tasks are often very simple when done in a timely way, but they can become huge problems if neglected or put off. We have found scheduling tables to be quite valuable for staying organized as well as for planning time and budgets. This also is a quick and easy way to communicate your intentions to others, whether family members, clients, volunteers, or other stakeholders. An important part of most restoration and management plans is scheduling work, especially when it involves other people. Their involvement is more enthusiastic when they have a long-term vision of what will occur on a project.

As you get more sophisticated in your scheduling, the scheduling table and the simplified cycle table can be tied to base maps showing the units where work will be conducted. If results of monitoring and scheduling are recorded in a tabular format, particularly a computer spreadsheet, you will have a readily accessible record of the restoration history. Ideally, this should be supported with maps and photographs, which can all be digitized for ease of filing and reference. This also helps when adaptively changing the plan.

## Conclusions

Not everything you do in your restoration will be successful. Even with careful research, good planning, and good execution, weather can defeat your efforts. Drought, late frost, and gully-washing storms can be as devastating to ecological restoration as to a corn farmer. Far more than getting a good crop of corn established, putting an ecosystem back together involves countless species and so many more variables, many of which you cannot anticipate. This provides plenty of challenges but also wonderful surprises. Careful planning and execution increase the probability of succeeding, and the challenges make success all the sweeter. Where possible, start projects with modest investments and scale up as you gain knowledge and experience. Always, however, pay very careful attention to what nature has to teach.

# Chapter 5

# Monitoring Progress

*There is one outstandingly important fact regarding Spaceship Earth, and that is that no instruction book came with it.*

Buckminster Fuller

All of us monitor ourselves and our environments continually, subconsciously if not with intent. The temperature of our homes, the fuel in our cars, the pain in our joints—the list is endless. Monitoring is the way we determine if systems are functioning properly. Monitoring is no less critical for ecosystems that we are restoring or managing, and it should be done systematically and regularly. Regardless of how monitoring is done, it should follow a standard methodology that is repeatable. In this chapter, we describe how to develop and use a good monitoring program and describe some of the more commonly used methods.

In monitoring human health, a host of things from body temperature to urine chemistry is used. Aches, sleep patterns, and appetite are qualitatively evaluated. Even with accurate measurements, years of data from large numbers of people commonly are necessary to differentiate healthy conditions from those that are unhealthy. Ecosystems are much more complex, and there is no simple indicator of ecosystem health. Most things you can measure require baseline comparisons or trend lines to be very meaningful. Water quality or diversity, for example, means little unless you have considerable experience in similar ecosystems or have a long-term database, often requiring years of observation. Because of the complexity, multiple parameters and observations are desirable.

Most restoration projects involve landscapes and multiple ecosystems. There are dozens of different tasks, with progress toward goals often measured meaningfully in years. In contrast, it may be possible to restore the functionality of a stalled car by one simple task, such as replacing the battery. Whereas the problem with a stalled car usually has a single mechanical cause and a solution, ecosystem dysfunction rarely results from a single problem. Monitoring needs to provide essential information to enable you to make informed decisions about restoration and therefore typically involves many components.

In ecological restoration it is best to try to simplify the complex cause-and-effect relationships to a hoped-for set of actions and responses, each with a beginning and end that can be evaluated. These actions should all contribute to the overall functionality of the ecosystem but collectively can never come close to dealing with all functions of the entire system. The complexity of the ecosystem is so great that one usually cannot know what all the problems are, much less which actions are most likely to improve the overall functionality. Ecosystem functionality results from interacting elements of an ecosystem: habitat and community structure, biodiversity, and biogeochemical cycles. As they interact, these processes create structural patterns (e.g., communities or cover types and their distribution over a landscape); products (e.g., clean water, soil organic matter, fruit and seeds, clean air, plant and animal biomass, etc.); and processes (e.g., groundwater infiltration, pollination, predation, succession, etc.). Evaluating the progress of a restoration project requires you to review and affirm the key elements of the restoration plan, usually annually. This involves asking simple questions:

- Has the vision been met?
- Have goals and objectives been achieved?
- Are there any new objectives, or objectives that need to be modified?
- Have the tasks been completed, and are additional treatments needed in any particular management unit?
- Do the monitoring results indicate success in each area or for each treatment?
- Overall, is the restoration moving the ecosystem toward greater sustainability?
- Are there units where monitoring results indicate that the vision has not been met?
- Does the program plan require modification? Particularly, are there other specific tasks needed to achieve success?
- If the ecosystem conforms to the vision, as indicated by measurable goals, should maintenance tasks be started, at least in some units?

## Affirming Measurable Objectives

When reviewing a restoration project, begin with a review of the goals and objectives that were established for the project. If success has not been achieved in any goal or objective, this is the time to reappraise the restoration plan, goals, methods, and criteria used to evaluate success. It is not uncommon for changes to be needed, especially in the first few years of a project. In even the simplest restoration project, something will be learned. As information is gained, it often leads to revised goals and objectives. The review process is the key to capturing that information and coupling it to revisions in the project plan. Ecological restoration is best viewed as adaptive management. Changes in the restoration program, however, should not be made in haste. Changes must be merited and suggested by monitoring, the results of measures and observations.

During the review process, you may find a few goals that are unequivocally achieved. There may be uncertainty or even disagreement among participants or others on what needs to be done next. To limit uncertainty as much as is practical, develop measurements or observations that objectively relate to each specific objective. After data are evaluated and discussed, ideally there will be consensus on whether goals and objectives are being met. If not, then it is likely that either the goals and objectives need to be refined, or different measurements are needed, or inadequate time has passed to see measurable results. In this case, more time and patience may be required. It is common for goals and objectives to be refined and redefined, adaptively, over the years. This process also can help determine how much personal time and money are going to be required. Soul-searching is often necessary to determine how much you are prepared to invest, and what results you need to stay invested and motivated to continue.

Ecological monitoring, done properly, will provide important data about the effectiveness of the restoration and management program. It requires that the response of the restored communities be checked regularly. This is usually done by measuring ecological indicators of community recovery, which typically are based on healthy reference areas. Effectiveness is judged against the goals and objectives of the project design. The results will direct the management activities, including retreatment, if necessary, for the upcoming year. Although measurable data are important, observations and photographs are useful and are commonly used to supplement measurement data. With either, it is important to keep careful records, including dates, locations, weather conditions, or any other variable that might influence the interpretation of changes being noted.

Monitoring also can be used to evaluate what would become of the land if no restoration treatments were initiated. Having a control area is sometimes the easiest way to understand the results of treatment or demonstrate the impact of various stressors. It is common to set aside control areas to determine if ecological health will return once stressors, such as livestock grazing, are reduced or removed. Typically, this involves a small cordoned-off area where baseline and follow-up measurements and observations are made to assess change. This is an especially important strategy where techniques such as prescribed burning are used. Having a control no-burn plot can very quickly and effectively demonstrate the effects of the burn treatment.

Monitoring should be simple and easy, yet provide information that shows the response and change in the land. Depending on the type of community and the time frame of restoration, monitoring may be assessed annually or on two- to five-year intervals. The simplest monitoring involves establishing locations where vegetation, soils, or some measure of hydrology can be repeatedly sampled before and after restoration has begun. Qualitative sampling can be as simple as using repeated photographs from permanent photographic stations. Monitoring can be more detailed, if necessary, and range from a tally of plant species in a defined area to estimates of the abundance of each species, usually with visual estimates of the percentage of ground each species covers. Plant species may be categorized (e.g., perennial, biennial, or annual; invasive or native; herbaceous or woody). Permanent plots can be marked with flags, metal (best if you are using prescribing burning treatments) or wooden stakes, or surveyed from reference locations, such as fence lines or posts along a fence line. Alternatively, global positioning system (GPS) is relatively inexpensive and accurate enough for most monitoring. Good record keeping is a must.

Monitoring faunal populations is more difficult because animals tend to be secretive and mobile. Surveys of breeding birds and frogs, however, can be both fun and valuable. There are simple to very sophisticated ways to monitor hydrology, infiltration, macroinvertebrates, and vertebrates. Water quality requires extensive baselines for comparison and often is extremely variable throughout a storm event and from one year to another. Nevertheless, such data may be necessary to assess important changes in the ecosystem. Monitoring should be kept as simple as possible but be sufficient to assess your progress toward goals and objectives. Ideally, monitoring should be something you and your partners or family enjoy.

## Basic Monitoring Techniques

Monitoring can directly or indirectly document changes on the land. Indirect methods are observations such as repeated photographs from the same permanent photographic location or measurements of water quality draining from the ecosystem. Direct monitoring typically involves measuring changes in the plant community or habitat. Changes in plant-community composition (the types of plant species present) and plant-community structure (e.g., the proportion of trees, shrubs, and herbaceous plants and their distribution across the landscape) are most common, but other measures such as the percentage of cover of each species or functional group, or frequency at which species occur in samples, are also used. It is especially useful to monitor the percentage of cover of nonnative invasives, or perhaps a particular invasive species that is seen as the primary challenge to restoration. Some common monitoring methods are described here.

## *Light Quality and Quantity*

Light quality and quantity determine what species can survive beneath other vegetation, but they indirectly indicate the density and nature of vegetation shading the point of observation. Light is extremely variable and therefore requires multiple measurements and careful control to be useful.

**Vegetation density.** If you are dealing with a shade-suppressed site, such as an overgrown savanna or prairie with weedy tree invasion, monitoring the release can be done with a visual target-monitoring system. One example is a two- by two-foot white board, blocked off into uniform squares. When viewed at different heights and distances, the percentage of the board that can be seen is an indication of how much vegetation is between the viewer and the board. This measurement can be quickly taken, and if done in the same spots each year, provides very useful indication of vegetation density.

**Light quantity.** Light meter readings can be taken using a camera and f-stop levels or a light meter. More information can be obtained if light readings are taken at different levels. Common heights include ground level, one meter, and two meters above the ground. By taking parallel readings in the open, the percentage of reduction in light under a canopy can be easily calculated. Although there will be some variation according to cloudiness, using percentage reduction largely removes the hour-to-hour and day-to-day variation. Nevertheless, it is ideal to take light readings on clear middays, at about the same time of year. There are more sophisticated instruments that measure only the light waves used by photosynthesizing plants, but these instruments are expensive and seldom are used outside research applications. The technique is essentially the same.

## *Plant Community and Vegetation*

Vegetation is the most easily monitored component of ecosystem structure. It commonly is all that needs to be monitored.

**Timed meander search.** In this method, the observer wanders through an area compiling a plant-species list and noting one-minute intervals. The number of new species recorded is plotted on the *x* axis, and time on the *y* axis to generate a curve that reflects plant diversity. The greater the diversity, the longer it takes for the curve to level off. It is not necessary to know plant identities but is essential to recognize different species. Thus, the technique requires reasonable familiarity with the vegetation, or a very good eye. Species-time curves can be done for areas as small as a fraction of an acre, or as large as hundreds or thousands of acres. The larger the area, however, the longer the search and the more necessary skillful plant identification becomes.

**Photographs.** Repeated photography from permanent stations every year or two can document change very well. Often, photos are taken in each of the four cardinal directions at each station. Record date, camera setting, direction, and other pertinent information with each photo. Digital images can be stored electronically, of course. All photographs are most useful when fully documented.

**Community mapping.** Use an aerial photograph to map discernable communities or associations of plant species. Because historic aerial photos are available, we recommend this as part of the background for an ecosystem that is to be restored. Continuation will document changes into the future. Repeat mapping every few years. Other features, such as roads, trails, drainageways, and insect or disease outbreaks can also be mapped and compared.

**Invasive species.** Because invasive species are so important, it is common to quantify them. This can be done by sampling (described in more detail later) or by mapping. Mapping can sometimes be done using air photographs but is better if done through a combination of direct observation in the field and mapping from air photographs. Use of a good topographic map facilitates field mapping. The field technique involves use of reference points, compass directions, and distances to sketch approximate boundaries on a reference map that shows the reference points. Recent air photographs are ideal for this.

**Quantitative estimates of herbaceous vegetation.** A *quadrat* method is most commonly used to generate quantitative vegetation data. Although more time consuming, this method can provide the most detail about a plant community. A quadrat is a standard sampling frame with a fixed area. Circular frames have less edge and can be centered on a single point for repeated estimates, but square or rectangular frames are often used. For herbaceous vegetation, small quadrats, a square meter or less, are most common. A child's hula hoop is roughly a square meter. Hoops can easily be made of one-quarter-inch flexible plastic pipe, with a plug in one end that extends far enough to receive the other end and form a circle. Be sure to record the precise area of the sampling frame, because it will make a difference in how data are compared if different-sized frames are used in different years. Sampling is easiest if done systematically, at standard distances along compass lines, through the community you are sampling. With this method, one measures functional groups (e.g., grasses, legumes, woody seedlings, nonnatives) or estimates the cover of each species within each quadrat. The latter, of course, requires that species be recognized. In addition to cover, it is sometimes useful to do stem counts. Frequency, the percentage of quadrats in which a species or functional group occurs, provides estimates of distribution. High frequency indicates widespread occurrence. Frequency is quadrat-size-dependent, so consistent use of a standard-sized quadrat is important. From these data, you can generate species lists and richness and diversity measures.

**Quantitative estimates of woody vegetation.** It is often useful to quantify woody vegetation as well. This can be done using larger quadrats and a methodology similar to that already described. A common quadrat size is 1/100th acre (435.6 square feet, or a square of about 20 feet, 11 inches on each side). Using this approach, cover is usually not estimated, but tree diameters and stem numbers of woody plants are recorded by species. It is common to also tally snags or dead stems as well. Cover of woody plants is most easily done with a line-intercept method. A measuring tape of fifty or a hundred meters or feet is stretched along the ground on a compass bearing, or between stakes that may have been installed at the onset of the restoration project. Transects can be either random or systematically located, or a combination. For example, all transects might originate at random distances along a systematically located baseline. Either surveying the location or using stakes to mark each transect allows them to be repeatedly surveyed over subsequent years. An observer moves along the tape, noting where each canopy (edge of living branches) of each tree species intersects the tape and where that canopy ceases to intersect the tape. The total intercept of a species, divided by the length of the tape equals the percentage of cover of that species. Shrub or small tree intercept is typically recorded separately from mature tree intercept. It is possible to break down species into three or more size classes, usually based on diameter, and record the cover of each category.

## *Hydrology*

Simply measuring the depth of water in a pond, stream, lake, or wetland seems straightforward enough. It quickly gets complicated, however, when one considers timing, methods, sampling frequency, measurement during frozen conditions, and the actual quantity of water moving through a stream rather than depth at a point. Measure of water quality can be even more challenging. Quality is the chemistry and physical condition of the water. Quality is often indirectly examined using a biological integrity index. This index uses relative numbers of different macroinvertebrates or captured fish species as surrogates for actual water quality measures. Because invertebrates and fishes are very sensitive to water temperature and chemistry, especially dissolved oxygen, this indirect measure is quite meaningful. Effective capturing and sampling of the organisms and some modest level of identification skills are required but can be done without formal training.

While extremely useful, many quantitative measurements of hydrology or water quality require some training and often require specialized equipment. If you need to invest, the limitations and usefulness of each method and how to use instruments will be provided by the manufacturers of the equipment described later. We recommend talking with professionals who use that kind of equipment, however, to get their opinions and recommendations. We have found that even beginners can quickly learn to use soil moisture probes or calibrate stream gauges. The more important questions are what data are needed and which pieces of equipment are best for the job.

Most states have established regular stream monitoring using volunteers. Many offer workshops for those interested. This is a great way to learn what is needed for simple water quality and hydrology monitoring. Check with your local natural resources office for information for your state. For additional details, see product supply catalogs and the manufacturers' documentation for equipment. If you need more background on monitoring stream hydrology, there are many excellent books.[1]

**Stream discharge.** Staff gauges are metal rulers typically attached to something such as a fence post that can be pounded and secured into the stream bank. They are placed where the channel cross-section can be quantified. You simply read the measurement off the staff-gauge ruler that tells you the water depth. This can then be used with other measurements, described later, to compute the amount of water being discharged through the channel at that point in time.

Measuring the amount of water flowing past a point (usually expressed as cubic feet or cubic meters per second) requires some attention to detail. First, you need to do some simple surveying of the cross-section of the channel. This is done by stringing a level line or strong tape measure across the channel, from the top of one bank to the top of the opposite bank. Then, at increments (typically every one or two feet), measure from the line with a survey rod to the stream bottom and record that vertical distance. You should also record the water level and the time and date of the survey. With measurements in hand, draw a scale model on graph paper to determine the cross-sectional area of the stream. It should be expressed in square feet or square meters.

To estimate the quantity of water flowing through the channel, you need to know the velocity through the cross-section you have surveyed. This can be done with a stopwatch, or even a watch with a second hand. Toss in a floating object (Steve's instructor used a piece of orange peel, which remains visible even when it is just below the surface of murky water) at a measured distance (ten feet or three meters is adequate) upstream, and determine the number of seconds it takes to pass through the measured stream reach. You will need to repeat this several times and use the average velocity. Multiply the velocity in feet (or meters) per second times the cross-sectional area to obtain volume of water moving through the channel cross-section each second. As long as there are no major changes in the cross-section of the channel you have measured (bank slough-off, sedimentation in the channel, log jams, etc.), this is a good estimate of stream discharge. For future measurement, you need only note the height of water and the flow rate. In time, you will find a nearly constant relationship between height of water and flow rate, and thereafter you need only note the height of water to estimate discharge.

Although more costly, there are meters that can be installed along the margins of channels. With the cross-sectional area data for each, and with a few measurements of water speed at different water depths (called stages), you can download the water-level data from these recording stream gauges and compute the discharge at any water level within the range of those for which you recorded stream velocity. If long-term measurements are needed, this will save much time. Meters, however, range from $600 to several thousand dollars, and installation and use will require some homework or training. Most computations, however, are simple spreadsheet functions.

**Soil moisture.** As most ecosystems are restored, soils become better at retaining moisture and nutrients. Thus, soil moisture provides a good indication of change in ecosystem function. While easy to measure, soil moisture changes constantly through wetting and drying cycles related to storm patterns and seasons. Thus, if you want to monitor soil moisture, you may need to invest in equipment that allows quick and easy measurement. Direct-reading soil moisture meters range up to several hundred dollars. Less expensive, but also very convenient once they are installed, are fiberglass probes buried in the ground at various depths. Through an attached wire accessible at the ground surface you can read soil moisture in real time. This is especially useful where you want to document increase in soil moisture after disabling tiles or filling ditches as a part of wetland restoration.

To get more accurate estimates, you can calibrate electronic readings by gravimetric estimates. This is done by taking a wet soil sample, at the same depth as each probe, weighing it, and drying it in a 100°C (212°F) oven to the point where no further decrease in weight is observed. The percentage loss in weight is the percentage soil moisture of the wet sample. If you need only a few measurements, it is just as easy to forgo the probe and take samples as you need them. Drying time is usually two or three days, and the best data are obtained by sampling every few days through a complete drying cycle, from soil saturation to the driest condition the soil will reach in the field.

**Shallow groundwater monitoring.** If you want to document the response of shallow groundwater to restoration, there are very low-tech methods as well as automated data collection techniques to consider. Low-tech involves hand-augering a hole four- to six-feet deep. Pull the soil from the hole, and immediately insert a one- or two-inch-diameter (depending on the diameter of auger you used) PVC pipe into the hole. The pipe should have several small holes drilled within six inches of the bottom and have a cap glued over the inserted end. Attach a screw cap that can be easily removed on the top. Use a small retractable metal measuring tape, a string, and a fluorescent yellow or orange, water-soluble, Magic Marker pen. Before each measurement, run the string down the length of the measuring tape far enough to reach from the top of the PVC pipe to the bottom, and use the pen to color the string. After allowing the string to dry for a minute, carefully insert the zero end of the tape down the well to the bottom with the string attached. After ten to fifteen seconds, *slowly* pull up the measuring tape and read where on the string the water-soluble ink was dissolved away. This depth on the tape identifies the depth of the water. By subtraction, you can compute the distance from the top of the ground to the water surface. To document changes in groundwater depth associated with changes you are making on the landscape, groundwater measures like this can be taken in many locations over an entire season or over many years. Bear in mind, however, that climatic variation is a major factor, so be sure to keep careful records of precipitation and evaporation, available from an official weather station if one is nearby.

The high-tech method involves a continuous-reading electronic meter in the PVC pipe. Like the stream gauge, data can be downloaded to a spreadsheet that automates the computation of statistics. Useful averages that can be generated with this data include groundwater levels and minimum and maximum water levels, among others.

## *Faunal Monitoring*

Sampling fauna need not be a burden. Indeed, it is an activity that can excite children as well as adults about what is happening on the land. Although possibilities are nearly endless, we suggest here some relatively simple and easy techniques that nearly anyone can do.

**Spring breeding surveys.** Surveys for birds and some amphibians (frogs and toads) during breeding periods are relatively quick and easy. Surveys are conducted in early morning for birds, and at night or on cloudy and rainy days for frogs and toads. This involves standing quietly for five to ten minutes and recording the species heard or seen. Good tapes or CDs are available to help learn the songs for species in your area. Be sure to get recordings for your area, as they often are sold for different regions. Check with your local library; they may have the CDs you need or can help you find a source. Young people often have a very quick ear and can learn bird and frog and toad songs quicker than adults.

**Sampling fish and aquatic macroinvertebrates.** Surveys are usually done with nets. To sample fish, seine nets are pulled quickly through the water column over a set area (e.g., a hundred square feet). The captured fish are identified and enumerated, then released. For macroinvertebrates a small D-net is more often used, especially in streams. Nets are placed firmly against the bottom, and substrate just upstream of the net is kicked to dislodge organisms that float into the net. Captured organisms are then identified to order (mayflies, true flies, dragonflies, etc.) or family (not very difficult for aquatic macroinvertebrates) and enumerated. There are many excellent manuals to guide your identification.[2] A good resource for fish identification is the Environmental Protection Agency's (EPA) Freshwater Fish Identification, part of their *Biological Indicators of Ecosystem Health*, available online at http://www.epa.gov/bioindicators/html/fish_id.html. The invertebrate sample should be augmented with organisms taken from the bottom of stones or organic debris. Quantification is improved if sampling is limited to a set area, such as a sample unit of five square feet, and also if done by habitat type (e.g., comparing riffles, pools, runs, etc.) in streams; or in wetlands comparing the same vegetation communities. Be aware that the organisms collected will be different when sampling different habitats in steams or wetlands, or the same habitat at different times of the year. When comparing streams or sampling repeatedly over years, be sure to standardize the methods, including the date of sampling. Either ponds or streams can be sampled with these methods, although the D-net is used in a scooping manner against vegetation or organic debris along the shore when sampling still water.

**Snake boards.** Although not everyone enjoys handling snakes, many people are fascinated by them. An easy technique for sampling snakes and other creatures that like to hide under things is the use of boards that are placed randomly in a terrestrial setting. Boards should be of standard size (four-by-four-feet works well and can be made by cutting a sheet of plywood in half) and composition. Old plywood is better than new. Periodically, boards are turned to reveal creatures that may be hiding under them. **Please note that some snakes and other creatures such as spiders and**

**scorpions that hide under boards may be venomous. Use care in areas where they may be present.** A variety of invertebrates as well as snakes, salamanders, or other organisms are often present. When done systematically, compiled data can begin to reveal trends over the years. Boards can be left in place throughout the year, which improves the technique, but they should be sampled on established dates.

**Pitfall traps.** Many species, from wolf spiders to various ground beetles and salamanders, can be sampled with pitfall traps. These need not be elaborate, and can simply consist of a 28-ounce can with the top removed. The traps are placed carefully into holes of like size, so that they are flush with the soil surface. Attempt to keep vegetation or litter intact around the trap. Organisms usually are collected overnight, although traps can be left for 24 hours without harm to organisms. Do not leave traps unattended for longer than 24 hours or during rainy days. A few small holes in the bottom of cans, however, will prevent their filling with water and drowning captured animals.

**Bat surveys.** These are done at night using sensitive devices that can record the high-frequency pitch unique to each bat species. A computer program is then used to decipher the species by matching the recorded call with an online library of calls. These data are useful only to determine what species of bats are present and do not give reliable indications of quantity. Therefore, such surveys are more useful in areas where many bat species occur, especially in the tropics.

**Moth surveys.** Moths and some insects can be sampled at night by surrounding a portable fluorescent black light with white bed sheets. Drawn in by the light, the moths land on the sheets, and then are identified and enumerated. As with other samplings, it is not usually necessary to identify the organisms to species, but you need to recognize differences. If the methodology is standardized, including the date and time of sampling, these data can provide a good indication of changes taking place in an ecosystem.

**Sweep net for insects.** By far the most diversity in a terrestrial ecosystem, excluding microbes, is among the insects. With a bit of study and some reference books, it is possible to learn orders and even major families of insects. To sample insects in herbaceous vegetation, use a sweep net in wide arcs across the tops of the plants, standardizing the number of sweeps in each sample. Dump samples into labeled jars for later sorting and tabulating. Insects can be kept alive and released if processed promptly.

## Other Benefits of Monitoring

Monitoring is the basis for adaptive management, but beyond this, it guides changes required in restoration tasks. Monitoring, however, is not the only variable to consider. For example, you might implement prescribed burning but after a few years find that burning did not result in the changes you expected. Why? By framing questions and comparing the responses, you can begin to understand reasons for unexpected results. It is through monitoring that you gain these insights. During the remedial restoration phase, monitoring often drives the subsequent tasks and their scheduling.

Maintenance of ecosystems, with methods such as prescribed burning and treatment of invasive species, often becomes routine. This also means routine monitoring is needed to assess conditions in the ecosystem. Indeed, it may be dangerous to get into a routine that is not based on monitoring. For example, without monitoring, we have seen many landowners use prescribed burning about the same time annually or every few years, perhaps resulting in loss of some species, or at least failure to recruit species that might have responded to fire in a different season.

Monitoring should always involve revisiting goals and objectives, and perhaps even the reference areas, to be certain that the performance measures you have designed are realistic. Perhaps objectives no longer seem realistic, or goals have proven to be impossibly demanding on time and resources. When the schedule is too aggressive and the land is not responding as you hoped, nature is telling you something. Pay attention.

## Is There Ever an Endpoint?

Ecological restoration provides tremendous satisfaction, but at some point your resources and energy may be exceeded. Even ongoing maintenance can become demanding. Perhaps the cacophony of harmonizing toads, and competing peepers and tree frogs, singing the joy of their new-found habitats, is sufficient to keep you going. The monitoring process allows you to adaptively reconcile success with available energy and dollars. You should not become a slave to a plan, however, but look for ways to continue as a partner with nature to maintain what you can. Fortunately, successful restoration approaches the point where ecosystem largely maintains itself and your investments in time and dollars can be greatly reduced. The aim is to persist in your efforts until you reach that point. But, don't be fooled into thinking your ecosystems can ever become totally self-sustaining. In our experience only in the remotest of landscapes, where land is still connected to the larger ecosystem processes, are self-sustaining ecosystems even possible. Continue monitoring to keep your finger on the pulse of the ecosystem and be prepared to step up at any time management work is needed. Most ecosystems cannot become completely restored, at least to the point where they can function without any maintenance. Realistically, ecosystem restoration is a commitment forever, but hopefully the rewards exceed the effort.

## Conclusions

Evaluating the response of the land to restoration treatments will guide succeeding restoration decisions. If you have developed clear goals and objectives, monitoring will inform you of when real progress is being made. Through adaptive management, you continue to pursue your vision and refine strategies to become more effective. Being open to change is essential, but this should always be done in a structured way with a clear justification.

A realistic vision begins with your imagination and understanding, but as you learn through the monitoring process, it will evolve with your understanding of the land, as will the energy and resources you have available to invest in restoration. Uncertainty is part of every adventure, including ecological restoration. Especially in altered landscapes, the path of restoration and management is long, probably without a true endpoint, but the effort of early years eventually levels off, allowing more time to reflect and enjoy your accomplishments.

# PART II

# Applying Restoration to Different Types of Ecosystems

In part 1, we developed a systematic way to explore land and understand its ecological context and history. Then, we described how to develop a restoration plan, with goals and measurable objectives, and how to implement it. To facilitate understanding, we used the restoration of Stone Prairie Farm to illustrate the ten-step process. We now want to examine other types of ecosystems across North America, and explain how the basic process applies. In each case, we examine the ecological and historic context for biome or ecosystem type, the primary stressors impacting it, and practices that have proven useful in restoration.

The same step-by-step process developed in part 1 is used in ecosystem restoration in all types of ecosystems, although we do not repeat the specifics with each example we describe. The process always begins with investigating the ecosystem, how it works and what stressors are impacting its functionality. We point out documented stressors that have impacted and, in most cases, continue to impact the ecosystems. The same stressors are likely to be present in your project(s). Because it is impossible to examine every specific type of ecosystem, we use a broad-brush approach, with different types of ecosystems in the major biomes of North America. The aim is always to improve ecological functions.

From backyards to forests, from deserts to wetlands and streams, the framework for developing a restoration plan is the same. The differences are in structure and composition, which determines both services these systems provide and their sensitivities to different stressors. Even if your interest is primarily with one type of ecosystem, we believe some familiarity with other ecosystems will increase your understanding of ecosystem restoration. The variations, however, are endless. Each ecosystem is unique. Once you grasp the basics, you begin to appreciate the similarities as well as the differences. We cannot guarantee that you will always be successful, but we can guarantee that ecological restoration work is intellectually and physically demanding as well as spiritually rewarding.

# Chapter 6

# Grassland Restoration

*The wealth of the tall grass prairie was its undoing.*

John Madson

Mention "Midwest" to average Americans, and they likely will envision farmsteads scattered among fields of hay and grain, and maybe a pasture with spotted cows. Likewise, "West" conjures images of waving fields of wheat, or more likely, cattle and horses scattered across a vast landscape, perhaps a desert, or mountains. Few can now envision the vast reaches of native grasslands that once dominated these landscapes. *Prairie* refers to ecosystems dominated by grasses and forbs, with few if any trees, and is most often applied to the more robust eastern grasslands. *Rangeland* is more often used in reference to the short-grass prairies of the West, implying the widespread use of this steppe country for grazing. *Grassland,* therefore, is a general term that seems to be most appropriate given the dominance of grasses throughout these ecosystems.

Major grasslands also occur in South America, southern Africa, and in a broad, irregular belt from the Mediterranean to eastern China. Although the species vary, grassland ecosystems are remarkably similar throughout the world. All have been exploited for agriculture, especially grazing of livestock. We will focus on North American grasslands, although restoration techniques are largely the same elsewhere.

## Historic Grasslands in North America

From Indiana to the eastern slopes of the Sierra Nevada Mountains in the United States, north into Saskatchewan, Canada, millions of acres of grassland ecosystems have disappeared. They have become farm fields and rangeland. Each year, fewer remnants of healthy grasslands remain.

The grassland landscapes historically encompassed extensive wetlands and fragments of forested ecosystems along rivers and wetlands, where wildfires were less frequent. In North America, prairie outliers occurred as far east as western Ohio, with fragments on south-facing slopes and coarse-soiled ridges, often among oak and hickory savannas. Trees were not common in the grasslands, except along the margins where grassland species gave way to conifer forests in the West and North, pinyon-juniper in the Southwest, and oak savannas and deciduous forests in the East.

Wildfire, summer drought, herbivory, and dehydration from winter exposure and ever-present wind favored grasses and forbs over woody vegetation. Fires resulted from lightning and from Native Americans who ignited fires to drive game and clear vegetation to improve visibility and air movement, and perhaps to reduce biting insects. Grasses are better adapted to limited moisture and repeated grazing and fire. Where trees and shrubs did occur, especially in the eastern grasslands, they were repeatedly top-killed by fires, and buds were often killed by winter desiccation or browsed by elk and deer. These stunted trees, called *grubs*, could survive for decades. When fires became limited by development or climate shifts, grubs quickly grew to tree form and then were reduced again to grubs when conditions favored fire.

Throughout most grasslands, evaporation exceeds precipitation. Precipitation and evaporation rates follow inverse gradients from east to west, following the rain shadow of the Rocky Mountains. For example, annual precipitation decreases from thirty-seven inches at Kansas City, Missouri, to thirteen inches at Greeley, Colorado. Precipitation patterns also vary from south to north. Calgary, Alberta, Canada, for example, almost directly north of Greeley, receives an average of sixteen inches. Evaporation also varies with these gradients, with lower evaporation coinciding with locations receiving higher precipitation. Seasonal differences affect ecosystem functions. For example, plants become dormant during the cold winters in the north, while in the south, cool-season grasses, many of which are weedy, can grow through the winter.

North American grasslands are divided into tallgrass, mixed-grass, and short-grass prairie (fig. 6.1). Tallgrass prairie formed a narrow band, interspersed with oak savannas, in the eastern climatic zone where precipitation is greatest and evaporation lowest. Tallgrass prairie is an extraordinarily productive ecosystem and, as a result of annual turnover of extensive fibrous roots, gave rise to soils especially high in organic matter.

Short-grass prairie occupied the climatic zone closest to the Rocky Mountains, extending to western Dakotas, Nebraska, and Kansas. Mixed-grass prairie occurred between tallgrass and short-grass prairies. The composition of grasslands varied de-

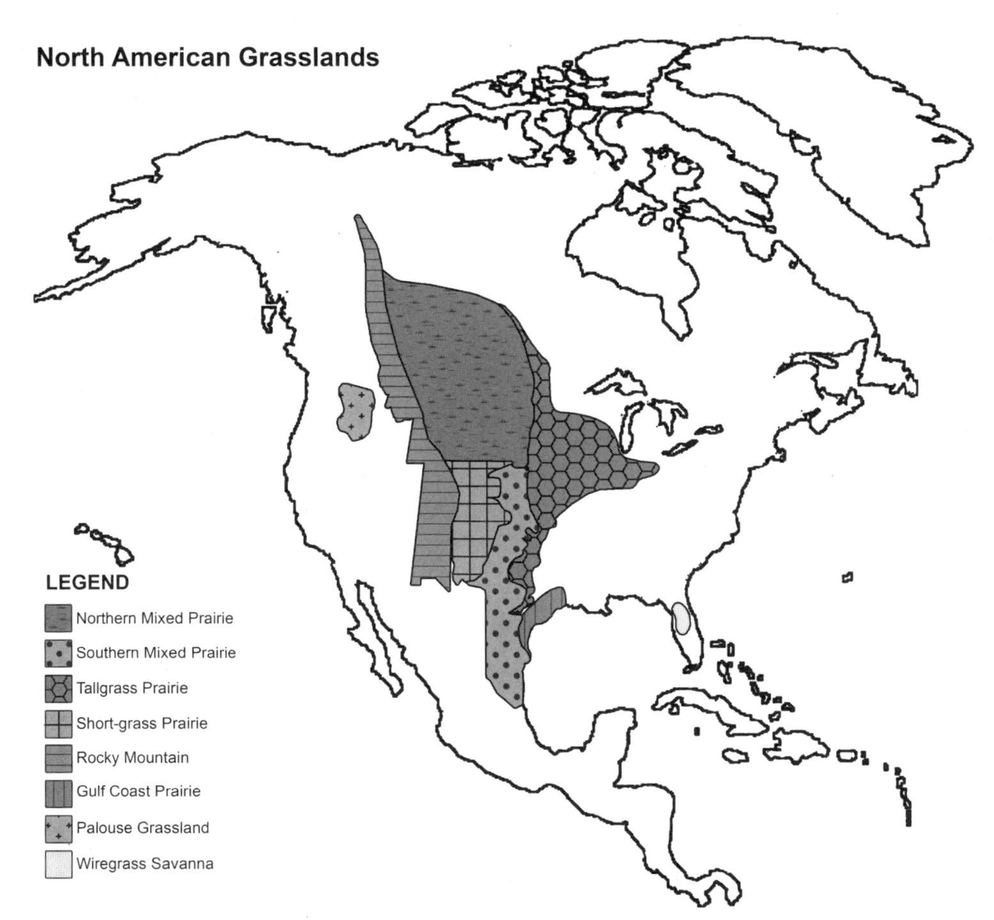


FIGURE 6.1 Grasslands of North America. A range of grassland types and transitions between types are shown. The major types are the focus of descriptions in this chapter.

pending on slope, aspect, and soil characteristics. Although conditions become increasingly drier toward the west, fire was more important where the productivity was highest, in tallgrass prairie. There, fuel loads were greater and fires were hotter and more frequent. Historically, short-grass prairie occupied about 615,000 $km^2$, mixed-grass prairie, 565,000 $km^2$, and tallgrass prairie covered 570,000 $km^2$.

Grasslands primarily developed in soils comprising loess, windblown silt, and mineral-rich glacial till. The soils associated with tallgrass and mixed-grass prairie are called *mollisols*. Mollisols have deep (60–80 cm), well-developed topsoils. Their high organic matter and excellent structure, together with a favorable climate, result in their being the most productive agricultural soils in the world. These prairie soils became the "bread basket" of North America.

Some mixed-grass prairies occupy outwash sediments carried by water from eroding uplands. Soils that developed on these finer-textured deposits are called *alfisols.* Alfisols develop more commonly under forested vegetation, but those that form with grasslands can be productive, depending on moisture availability.

Farther west, in short-grass prairies and deserts, much lower plant productivity results in soils called *aridosols.* There, evaporation greatly exceeds precipitation, and evaporative wicking draws mineral salts up into the soil profile where they accumulate, the depth depending on the deficit of precipitation. The accumulation of minerals (caliche) impedes root penetration, further limiting moisture availability. Consequently, aridosols are poor agricultural soils unless very well irrigated. Because vegetation is more sparse, wind and water erosion often are severe.

## Grassland Fauna

The historic abundance and diversity of the grassland wildlife is legendary. The journals of Lewis and Clark describe the abundance of species such as prairie chickens, geese, quail, bison, deer, and elk in the vast, remarkably diverse, grasslands.[1] Bison, in particular, were an iconic species with populations estimated at forty to sixty million animals that traveled in huge herds. Elk and pronghorn were also widespread, the latter being restricted to mixed- and short-grass prairies. Five species of prairie dogs, jackrabbits, ground squirrels, gophers, and voles were among the more common herbivore species. Prairie dogs are a *keystone* species. Keystone organisms have a prevailing influence on how habitat is used by others. Many other species, such as burrowing owls and snakes, use prairie dog burrows, and prairie dogs are prey for a wide variety of carnivores. Grasslands were also rich with insect herbivores including grasshoppers, butterflies, beetles, and ants, among others. Diverse bird populations including meadowlarks, various sparrows, horned lark, prairie chickens, upland plovers, and the species associated with shrubs and trees following drainageways also were present. In associated wetlands, such as along the Platte River in Nebraska and prairie potholes in the upper Midwest of the United States and Canada, an abundance of migratory waterfowl and shorebirds occurred. Wetlands were more diverse when surrounded by healthy grasslands, in contrast to wheat or corn fields that now dominate the landscape.

## Vegetation

The vegetation in each grassland type is different, although species overlap. The dominant grasses in the tallgrass prairie are big bluestem, little bluestem, Indian grass, and switchgrass. All also occur in other types of grassland. Cordgrass also is found in low, wet areas in all three grassland types. Needle-and-thread grass and western wheatgrass are more common in mixed-grass prairies, while short-grass prairies are dominated by blue gramma and buffalo grass. The abundant forb species reflect minor variations in topography and soil patterns, usually in response to moisture availability. Some species reach greatest abundance on warmer and dryer south-facing slopes, whereas others are more common on north-facing slopes or in draws where moisture is more available.

Species also are distributed along the east-west, north-south, and elevational climatic gradients. For example, Indian grass and big bluestem are more tolerant of higher heat and are more abundant farther south, whereas species such as western wheatgrass and many sedges have a competitive advantage farther north and at higher elevations. Regardless of species, however, grassland productivity increases with available moisture and length of the growing season.

## Fire

Importance of fire parallels productivity. Higher productivity means more combustible fuel and hotter, more frequent fire. In the tallgrass prairie region, there are accounts by early settlers of approaching wildfires, described as the "roar of an oncoming train." Fires driven by strong winds even jumped large rivers. Pioneers quickly learned to use fire as a tool, backfiring to protect their houses, barns, and livestock from approaching fires.

Short- and mixed-grass prairies also burned, but not with the intensity of tallgrass prairie fire. Fire and grazing are linked. In the short-grass prairie, fire suppression and overgrazing now make fire a rare occurrence, although the ecological importance of fire also varied prior to settlement. Fire was important in the evolution of grassland ecosystems throughout the world. Not surprising, therefore, fire is an essential tool in the restoration of grasslands.

## Stressors

With European settlement, big shifts occurred. Fires were suppressed, browsing game animals were replaced by cattle and sheep, and overgrazing became more common. As fire decreased, trees and shrubs expanded from the protected margins and spread into the grasslands. The invention of the steel plow by John Deere in 1837 resulted in rapid expansion of agriculture. Reduced fire, loss of grazing and browsing fauna, fragmentation, and agriculture took their toll, and nonnative weeds further traumatized grassland ecosystems.

Now only about 1 percent of the original tallgrass prairie and 50 percent of the short-grass prairie remain uncultivated. However, even where prairie remains, it has been greatly diminished by overgrazing and nonnative species. Healthy grassland remains primarily on infertile, sandy soils, such as in Nebraska's Sand Hills and the flint hills in Kansas, Missouri, and Oklahoma, where agriculture and land development have not been aggressive.

The dramatic loss of the grasslands happened with remarkably little notice. The degradation of remaining fragments also attracted little attention, probably because most people could not recognize changes in quality, health, and biodiversity of prairie vegetation.

Early settlers in North American were perplexed by grasslands. If the land did not grow trees, it initially was assumed to be infertile. The heavy sod itself was further impediment to cultivation. That changed with the introduction of the steel plow, which quickly exposed the hidden fertility beneath the tallgrass prairie. Extensive tilling of grasslands in the United States and Canada resulted in severe wind erosion and the Dust Bowl years of 1930 to 1936. This experience also led to the beginning of soil conservation.

The key stressors to grassland ecosystems are

- Fragmentation. Especially in tallgrass and mixed-grass prairie, fragmentation is damaging to wildlife but also results in genetic isolation of invertebrates and plants.
- Fire suppression. Regular fire is essential, especially for tallgrass prairie.
- Invasive species. Both nonnative and native invasive species have spread into grasslands, primarily as a result of suppressed fire and overgrazing. Several aggressive nonnative weeds, especially cheatgrass and leafy spurge, have become widespread. Woody vegetation also spreads into grasslands in the absence of fire and is a particular problem in the ecotones.
- Overgrazing. Vast, but widely scattered herds of bison have been replaced with more concentrated herds of cattle and sheep. Bison roamed, often grazing for only a day or two in any location, whereas livestock usually are restricted.
- Change in hydrology. Especially in the mixed-grass and tallgrass ecosystems, where more intensive agriculture is practiced, hydrology has been greatly altered by drainage ditches and tiles.
- Haying. Both mixed-grass and tallgrass prairies, along with associated wetlands, are commonly harvested for feeding livestock. Cutting the vegetation at the peak of its development, usually early to midsummer, is especially disruptive to the reproduction of birds and other wildlife. Seed production of many

prairie species is not completed until late summer, and this practice reduces the reproductive potential of vegetation.

- Fencing. Fences disrupt the movement of larger wildlife species, such as pronghorn, but also confine livestock, leading to overgrazing.
- Disease. Several diseases affecting wildlife have been introduced, such as canine distemper and brucellosis. The canine distemper virus nearly led to the extinction of the black-footed ferret and is a serious problem with wildlife the world over. Brucellosis was probably introduced to North America by imported cattle, but it now infects both bison and elk.
- Nitrogen enrichment. Atmospheric transfer of nitrogen, primarily from agricultural fertilizer application, can upset the nitrogen balance of natural ecosystems. Systems enriched with nitrogen become more susceptible to invasion by nonnative plants.
- Erosion and sedimentation. Many grassland remnants suffer from erosion and sedimentation, especially those adjacent to tillage or subjected to overgrazing. Both erosion and sedimentation alter soils in ways that are deleterious to native species and impair wetlands and waterways.

The greatest decline in grassland species and communities has been in the tallgrass prairie and the least in the short-grass prairie. Extensive conversion of mixed-grass and tallgrass prairie to cropland resulted in extreme habitat fragmentation. Larger animals such as wolves and bison require extensive unbroken tracts of grassland, but even many invertebrates are isolated in small fragments of prairie. Grassland birds and butterflies have declined 60 percent to over 90 percent since settlement. Recently, declines in amphibians are being widely documented, and many other vertebrates are threatened. Endangered species include black-tailed prairie dog, black-footed ferret, and western prairie fringed orchid.

Cultivation and overgrazing create opportunities for nonnative species such as ring-necked pheasant, which has contributed to the decline of the greater prairie chicken. Some introduced plant species also are serious problems. Crested wheatgrass, for example, widely advocated for improved winter forage, often develops monocultures on western rangeland. Leafy spurge is an aggressive perennial weed accidentally introduced into North America from eastern Europe in the nineteenth century. It currently infests over two million acres in the United States.

The decline in short- and mixed-grass prairie, including spread of invasive species, is directly linked to the beef industry. Overgrazing, erosion, reduction of soil organic matter, and fencing are associated with livestock. New models have been proposed for management of cattle to emulate the grazing patterns and behavior of bison with a rotation of grazing intensity. Rotational grazing is designed to better match grazing with growth response of the forage species, and stimulate higher yields with less damage.

Riparian areas occupy about 1 percent of the grasslands yet harbor a disproportionate amount of plant and animal diversity. Hydrologic modifications have favored invasive species such as Russian olive and salt cedar. Even the largest rivers in the West have been altered because of runoff from upland grasslands in their watersheds. (See chapter 9 for further discussion of riparian restoration.)

## Market and Policy Changes

While it is likely that agriculture will continue to dominate the Great Plains, changes are occurring in the marketplace that potentially will greatly affect grassland restoration. Preference for healthier foods and a concomitant shift from corn-fattened beef to grass-fed beef will favor restoration of grasslands. Limited water and rising costs of intensive agriculture also are shifting the economic advantage toward grass-fed meat. Policies to address climate change will add additional economic incentives for restoring grasslands. The Conservation Reserve (CRP) and Wetland Reserve programs of the U.S. Department of Agriculture (USDA) are likely to be revitalized, but most public officials will need help understanding the benefits of restoration of grassland ecosystems.

Soil organic matter is second only to the oceans as a sink for carbon in the global ecosystem. No ecosystem is more effective at putting carbon into the soil than temperate grasslands. Moreover, much attention is being directed to biofuels—switchgrass, for example. But diverse tallgrass prairie ecosystems are more sustainable and, over time, more efficient than a monoculture. Market forces may increasingly lead to carbon "farming," replacing corn, soybeans, and feedlots with corresponding improvements in both the environment and human health. Unlike other commodities, carbon farming has the added benefit of delivery of the "crop," carbon credits, anywhere in the world, on paper, while the carbon enriches the soil, further increasing its productivity and ability to absorb water.

Historically, USDA advised planting reserved acres with nonnatives such as smooth brome and western wheatgrass. They are now accepting and even encouraging farmers to restore native prairie on reserved acres. Monocultures are nearly devoid of bird life, whereas native grasslands are remarkably diverse. A high-quality native prairie may have two hundred to four hundred vascular plant species and a dozen bird species in a patch of no more than five acres. Prairie restorations often start with five to ten species of plants, mostly grasses and a few forbs, but even that is a far superior habitat for wildlife. Over time, restored prairies are invaded by native goldenrods, asters, and other plants, many of which have wind-disseminated seeds commonly found in low abundance around the edges of disturbed lands.

## Restoration Practices

There is a significant and growing enthusiasm for grassland restoration, above and beyond market- and policy-driven incentives. Even homeowners with small urban lots are planting native prairie as an alternative to formal lawnscapes. In the Midwest, many corporations that would otherwise be maintaining large lawns at considerable expense are converting them to prairie plantings. Parks that have been maintained for years as formal lawns are converting bluegrass to tallgrass prairie, acre by acre, with a corresponding decrease in operational costs and increased benefits to wildlife and environmental quality. Aside from the functional benefits, people are discovering the beauty of diverse native grasslands.

Another surprisingly large movement is coming from people who have purchased land and moved back to the country in lieu of urban lifestyles, or have a second home in the country. These property owners are usually more interested in the quality of their environment than profits from wheat, corn, and cattle. An entire prairie-restoration industry has emerged, selling knowledge, plants, seeds, and services such as restoration and prescribed burning.

Whether a grassland restoration is successful or not often depends on the landowner's expectations. Landowners often believe that once the seeds are planted, they should spring from the ground and instantly create the aesthetic benefits and vibrant life found in a native prairie. This, of course, is a false expectation, probably stemming from years of experience with annuals in agricultural fields or flower gardens. Nor is a native grassland like a new lawn where sod can be laid or seeds sown with satisfying results in hours or, at most, a couple of weeks under the sprinkler. Prairie plants invest the first few years in developing strong and wide-ranging root systems with rather diminutive aboveground development (fig. 6.2). Even those that develop most rapidly will not demonstrate robust foliage and flowering until at least the second or third year. Because of the time required for prairies to establish and develop, it is important to help landowners understand the process and develop reasonable expectations.

### *Mixed-grass and Short-grass Prairie Restoration*

A first approach for restoring any ecosystem is to mitigate the stressors, hopefully allowing the ecosystem to heal itself. The restoration of mixed- and short-grass prairie

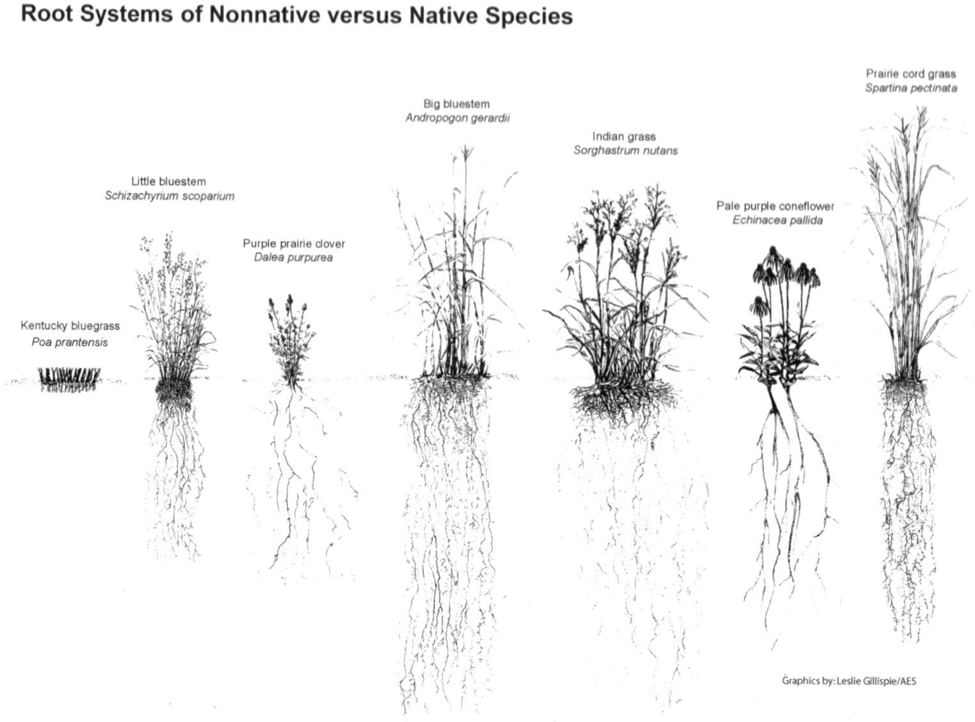

**Root Systems of Nonnative versus Native Species**

FIGURE 6.2 Root architecture and complexity of natives versus nonnative grasses.

restoration may differ significantly from tallgrass prairie recovery because of differences in primary stressors. Unlike the conversion of a former cornfield to tallgrass prairie, where agricultural weeds will be a major hurdle, in short- and mixed-grassland restorations the native grasses are still more likely to be present and weeds less of a problem. However, most short- and mixed-grass prairies have been badly overgrazed. Depending on conditions, restoration will begin by eliminating or reducing livestock grazing to allow the grass and associated vegetation to recover. It may also involve reseeding disturbed areas or diversifying existing vegetation by reintroducing native forbs and grasses. In deciding what to introduce, pay particular attention to plants that are typically reduced by livestock grazing. Also of great importance is stabilizing bare soil areas to secure them against the risk of invasive species.

Control of many invasives is quite problematic in vast western rangeland. Some biological control measures, such as the leafy spurge flea beetle, are showing good promise, but more time is needed to fully evaluate the efficacy of most. Control of cheatgrass is a particular problem. Because it is an annual and has short-lived seeds lasting only a year or two, reducing seed set appears to be the most promising strategy. This is being done with herbicides, use of prescribed fire, mowing, and intensification of livestock grazing at the green-leaf stage to slow the maturation of the plant. Herbicide use, of course, is a problem, because those that kill cheatgrass also kill desirable prairie species. There are many more details to be learned on the management of these and other invasive plants in grassland restoration.

## *Tallgrass Prairie Restoration*

Tallgrass prairie restoration initially involves controlling the growth, or at least the dominance of aggressive, nonnative perennial grasses and forbs. Proper site preparation is essential, and often determines success or failure of restoration. In former agricultural fields, thousands (sometimes hundreds of thousands) of weed seeds will have accumulated in each square foot of soil surface. While both desirable native species as well as weedy species usually will be represented, years of agriculture will have resulted in the loss of most of the native species and accumulation of seeds of weedy species.

Initial control of weedy species is typically done by tillage with or without herbicide treatments prior to seeding. Often conventional cropping of land for a few years, with conventional tillage and herbicide treatments can reduce nonnative competition in advance of prairie establishment. Make sure desirable native species are insufficiently represented in the site before using herbicide. If herbicide is needed, treat the existing vegetation with glyphosate or another relatively nontoxic herbicide. If converting areas in which smooth brome or western wheatgrass is dominant, two treatments are often undertaken, the second within a few weeks of the first. Often, the existing vegetation can be mowed and harvested for bedding in late summer and then sprayed after it greens up again before frost. A second herbicide treatment may be needed in the spring if the first treatment does not completely kill the perennial vegetation. If the soil has a lot of viable weed seed, you may need to cultivate between treatments to stimulate the germination of weeds. Alternatively, a year or two of glyphosate-resistant soybeans or corn will eliminate perennial grasses, such as smooth brome, and reduce the weed seeds. If cultivation is not required, prescribed burning may be necessary to eliminate the dead thatch. A no-till drill can be used to insert seeds directly through the dead sod.

Once the weeds are under control, the native plant seeds are introduced by drilling or broadcasting, sometimes with cover crops of annual, biennial, or nonpersistent perennial species. Additional plant species can be introduced after the initial native prairie is established, either by seeding or planting, usually after a prescribed fire. Anecdotal evidence indicates that many less-common species may not become evident for a decade or more after being introduced, so patience is important.

## *Choice of Species*

An important decision is which species to plant. We recommend a reasonable mix of the primary forbs and grasses native in local prairies on similar sites (reference areas). Prairie plantings can be diversified over time with locally collected or purchased species. Either seeds or plugs can be used to augment what is present. Be cautious about including too much of particularly aggressive native species. They will impede establishment of other native plant species and reduce the overall diversity.

Especially when restoring mixed- and short-grass prairie, be cautious of recommendations promoted by agricultural interests, often from state or federal agencies. Avoid recommendations focused on grasses that do not include a diversity of other plant species. If you are not interested in creating a near monoculture, seek other advice for planting mixtures. The typical difference is that the restored prairie has a balance of grasses and forbs, rather than a dominance of one or two species. Even in tallgrass prairie, too much switchgrass or Indian grass can make introduction of other forbs and grasses difficult. After the first two or three years, watch for small patches of bare soil where new species can be added to increase diversity. Too often an agricultural model is used, where seeding is done with one or two species sufficient to occupy every square foot of soil.

Attention to the source of seed is seldom a concern in agriculture. In prairie restoration, seed source is very important. Not all big bluestem or blue gramma is the same. Agency program specifications often fail to recognize the importance of using local seeds. In Illinois, for example, many prairie restorations were done with grass and forb seeds from Missouri, Kansas, Nebraska, or even New Mexico. You can anticipate that the more foreign the seed source, the more likely the species will suffer from insect or disease problems, or falter during stressful conditions. The result is often loss of diversity and functionality of the ecosystems. If for no reason other than being genetically prudent, we recommend the use of local genetic sources of seeds and plants. In theory, they will be the best adapted to your conditions of climate, soils, pollinators, predation, and diseases. There is good evidence, for example, that some imported races will flower out of synchrony with natives and not effectively set seed, or in the case of insect-pollinated plants, not flower when native pollinators are available.

## *Caring for New Prairie*

Agricultural weeds are mostly annuals, which means they tend to grow early, fast, and tall. On the other hand, most native prairie plants are either biennials or perennials. Biennials typically form a low-growing rosette the first year and flower the second year. Perennials, because they depend on belowground structures for so much of their existence, invest large amounts of time and energy in root production and may show very little aboveground growth in the beginning. A typical native prairie perennial may have ten to thirty times as much root as is shown in aboveground foliage. For example, a mature lead plant is about two feet tall. While it is attractive up close, its ecological beauty is below the surface—its roots can reach as deep as thirty feet! Extensive studies by John Weaver and his colleagues during the 1930s revealed the remarkable root systems of native grassland plants and showed their very elaborate network in grasslands soils.[2]

Because of the contrasting strategies of rapid growth and short lives of weeds versus slow growth and longer lives of prairie plants, most people will see only a field of weeds for several years after prairie, especially tallgrass prairie, is started. Be patient. The prairie plants are probably there, and you can use the weeds to favor the prairie.

During the first growing season, when the vegetation reaches about one or two feet, mow to a height of six inches. This prevents the weeds from producing seeds, and the prairie seedlings are too short to be injured. We also recommend not watering or fertilizing because that primarily benefits weedy species. Native plants are adapted to natural conditions and require no additional watering or fertilizer.

If the site was properly prepared, the weed seeds in the soil have been greatly reduced, or undisturbed if no-till planting is done through dead sod. In the second and third years, the native prairie plants, with their well-established root systems, now begin to allocate a greater portion of their energy to aboveground plant tissue. A succession of annual weeds to perennial prairie has begun. Remember, this is not an all-or-nothing process, and some weed species can persist for years. Prairie plants, with their increased production of aboveground structures and their superior belowground root systems, will gradually outcompete and replace the weeds. Expect some prairie plants to flower in the second year.

Since soil disturbance is essential for the weeds to continue to survive, resist the temptation to pull weeds by roots. Even the small area of soil disturbed by pulling a weed can let many seeds that are still in the soil emerge. Keeping the weeds mowed to prevent seeding is all the control needed in the first two or three years.

## *Prescribed Fire*

**Note: If you are not familiar with controlled burning, and the often-required certifications, licenses, and permits, please consult a professional.**

There are so many excellent resources on the proper use of prescribed fire that we include only a limited discussion here. If you plan to do your own burning, we urge that you first learn how to develop and implement a fire management plan. Pauly provides an informative booklet on this process, and we strongly recommend it for beginners.[3] Needless to say, anyone contemplating the use of fire as a management tool should have training in its use, paying considerable attention to the details associated with fire hazards and risks. Notifying local fire officials and getting burning permits are part of the fire planning and implementation process. Volunteering to assist others who have good experience with prescribed fire is a good place to start. The Nature Conservancy uses a lot of volunteer help and does an excellent job training their volunteers. There may be other organizations in your area that would welcome help and provide training and experience.

We want to emphasize a couple of points that might not be made by those from whom you learn. Many people take the order to burn an area too literally and leave no spot unburned. In contrast, we recommend starting the burn near the end of the day, or on days when the fire will be less aggressive. This is not only safer, it is better for the diversity of the ecosystem. In late afternoon and into the evening hours, as the relative humidity increases and temperatures and winds decrease, fire intensity decreases. Evening burns are more irregular over the landscape and contribute to the overall diversity of the plant and animal community responses, leaving partially burned areas, fire-blackened areas, and completely unburned biological islands. Over the years, with repeated burning, shifts in wind and conditions result in a shifting mosaic of fire effects, similar to natural fire dynamics. These results are completely different from what is produced in using fire to completely blacken the landscape. This difference in approaches to fire is all the more important when burning isolated fragments of prairie, where species have fewer places to escape fire, or from which they can reinvade afterward.

In restoration and management of grasslands, fire is applied to emulate evolutionary relationships established over the past ten thousand years. Objectives for the use of fire in grasslands include killing or suppressing undesirable woody plants; reduction of invasive species, primarily by maintaining healthy populations of grassland species; and increasing or maintaining the diversity of native plants and animals. Fire is especially critical in tallgrass prairie, where the accumulation of litter is associated with a reduction in the diversity of plants and some faunal groups. Fire is very useful in restoring short- and mixed-grass prairies, although grazing often removes much of the aboveground annual biomass in rangeland. Prescribed burning reduces old plant materials and stimulates new, more palatable growth. Fire also suppresses trees and shrubs and reduces competition from invasive species, nearly all of which are cool-season plants.

Timing of the burning is important. Cool-season grasses are detrimentally affected by late-spring or early-summer burning, which stimulates warm-season grasses. In the arid West, vigorous growth after fire sometimes does not occur because of unavailable moisture, whereas it is a reliable response in mixed-grass and tallgrass prairie.

The first prescribed fire after restoration should be planned for early spring at the beginning of the third year for tallgrass prairie, and as needed in mixed- or short-grass prairie. One fire, even a good one, will not rid your prairie of all weeds, but burning can reduce the need to mow, and increasingly favor prairie species over the weeds. Several successive years of fire may be necessary until weedy vegetation is largely eliminated. Generally, after the fourth year, prairie plants will be well on their way and it may be necessary to burn only every two or three years. The tallgrass prairie should become increasingly colorful as more and more of the forbs reach sufficient size and vigor to flower.

We are often asked if there are alternatives to burning. Many prairie restorations are in locations surrounded by homes or pine plantations or near major highways, where prescribed fire is more risky or problematic. Ecologically, there is no good alternative to fire. Prairie plants are adapted to fire. Burning removes accumulated litter and allelopathic chemicals, releases nutrients, neutralizes organic acids, and blackens the surface, which in turn causes the soil to warm faster in spring. Because prairie plants begin growth later in the spring than most perennial weeds, spring fires discourage weeds and favor prairie species. In addition, fire mechanically wears away or scarifies the often hard coat around the seeds of some native plants and encourages their germination. If burning is too intimidating or risky to do yourself, look for local professionals to do it for you. Also, do not neglect to involve and educate neighbors and other stakeholders, thereby increasing their tolerance for fire management. Mowing has little of the same effect, but it will reduce the growth of woody plants that have invaded, and if material that is mowed is removed for hay or bedding, accumulated and potentially combustible litter will also have been removed.

You may wonder if there ever will be a time when all the weeds are gone. On even the oldest restorations we have done, and those done by others that we have examined, we can say with certainty that the answer is no. Even minor soil disturbances such as ant mounds and animal dens provide sufficient habitat for some weeds. These weeds are not harmful as long as they are kept to a minimum with periodic fire. In fact, some native plants that might be considered weeds, such as evening primrose, are components of the prairie ecosystem, although discouraged in agricultural crops. Nor are all weedy species annuals. A common midwestern weed in prairie is Kentucky bluegrass, a persistent perennial. Such species are not easily removed or replaced through succession, competition, mowing, or fire. Indeed, they seem to fit into prairie management pretty comfortably. While these nonnatives may not be fully eliminated, good maintenance can reduce them to minor components within the prairie landscape. Monitoring will be essential to keep track of these invasives and make sure they are not increasing.

## *Control of Woody Invasives*

A common problem in grassland restoration is invasion of trees and shrubs. In many locations, the prairie may be reduced to wispy prairie plants struggling beneath the shade of trees and shrubs. Sometimes there are clumps of prairie grass and diminutive prairie forbs in the dense shade of pine plantations that were planted by landowners or professionals who failed to recognize the diminished prairie. More often, however, invasive trees or shrubs have invaded once-healthy prairies. Inevitably, this occurs as a result of cessation of wildfires, although dewatering of the landscape by agricultural drainage networks can contribute to tree and shrub invasion of wet prairie.

Because their extensive root systems are capable of storing much food reserves, many prairie plants can live for decades under dense shade. This brings us back to the ten-step restoration process, specifically site inventory and analysis (chapters 2 and 3). Early in restoration planning, as the site is inventoried, look carefully for suppressed native grassland species. If the site has not been severely altered by development or agriculture, many will be waiting to be liberated from suppression. Restoration of suppressed prairies requires patience, but it is rewarding.

Often, other invasive plants favored by the shade have become established under the woody vegetation. If you open the site quickly, these invasive grasses and forbs can be favored, resulting in a problem that might be even more challenging. The first step, therefore, is to learn what other plants are growing in the understory. Are there invasive buckthorn or honeysuckle shrubs? What about garlic mustard or burdock? If so, these should first be reduced or eliminated, before removing the overstory tree or shrub canopy. Depending on the invasive species, they might be reduced by hand pulling or spot treating with glyphosate. If prairie plants are under taller invasive grasses or forbs, mowing might favor their release. However it is done, the intent is to shift conditions to favor the prairie plants by providing them with more light and, when possible, exposing the vegetation to repeated fire. Usually, herbaceous weeds will be outcompeted by the long-lived, deeper-rooted, fire-adapted prairie plants, once they have adequate light to grow.

Fires were an important way that woody encroachment was controlled in the past, and prescribed burning can be one of the least expensive and most effective methods for reducing woody vegetation. Before prescribed burning is initiated, systematically go through the area and girdle and/or herbicide selected species. After the weedy herbaceous and sprouting woody species are under control, fire may be all that is necessary to release the prairie. A few years may be required for fire to eliminate woody plants, and some fire-adapted species will never be completely removed by fire alone. In tallgrass prairie, for example, patches of gray dogwood or American hazel are part of the native structure and should remain. Species that have not co-evolved with prairie, however, will generally be reduced or eliminated in time.

## Conclusions

Grasslands need to be restored for a range of ecological, social, and economic benefits. Climate change alone poises such a threat to our society, not to mention other inhabitants of the Earth, that no further incentive is needed for carbon sequestration wherever possible. Soils are one of the largest carbon sinks in the global ecosystem and none contribute to it better than temperate grasslands. The restoration of a prairie is not only a fun family project, but has far greater benefits than recycling plastic or maintaining a compost pile. Tallgrass ecosystems may become an important source of biofuel. Mixed- and short-grass ecosystems lend themselves well to carefully managed grazing. These benefits can be gained even as soils are protected and enriched with carbon, with corresponding benefits to watershed protection, wildlife habitat, aesthetics, and global climate.

## Chapter 7

# Forest Ecosystem Restoration

*When the oak is felled the whole forest echoes with its fall, but a hundred acorns are sown in silence by an unnoticed breeze.*

Thomas Carlyle

Forests provide many goods and services and the demand is growing. They also are important in watershed protection and have special aesthetic appeal. Moreover, studies now show that healthy forests have the potential to absorb much additional carbon dioxide. Nevertheless, it was not until near the close of the twentieth century that much attention was given to sustainable forest management, the meaning of which is still the subject of discussion. As we have become increasingly isolated from forests, our appreciation of them has increased, but our understanding of how these complex ecosystems function remains far from complete.

Emphasis in forest management historically was directed at the trees rather than the ecosystem. Cutting only the more valuable trees, called *high-grading*, was a standard practice and continues on much private land where logging occurs without the oversight of professional foresters. Only within the past twenty years has there been a slow shift, even among professional foresters, toward ecosystem management.[1] Managing the forest ecosystem shifts the focus from outputs, what the forest can provide, to sustainability. With this comes the realization that the myriad species, complex structures, and even disturbance regimes must be addressed in management.

The science of forestry had its roots in Europe early in the nineteenth century. However, the application of ecology to forestry received little attention, and it was only in the last two decades that the forest began to be viewed as an ecosystem. Even now, many foresters do not think ecosystem restoration has much to do with forestry. Forest restoration should be a consistent aspect of forest management, but it is seldom viewed as such. Typically, forest management focuses on optimizing goods and services, often with an aim toward sustainability, whereas forest restoration focuses on optimizing ecological health or integrity of the forest, which has come to be called *ecosystem management.*

Years of exploitation and lack of understanding, along with fire suppression and air pollution, have resulted in degradation of forests throughout all developed regions of the world. Only in remote areas have forests tended to remain in near-pristine condition, and even there, acid deposition, climate change, introduction of nonnative species, and disruption of presettlement disturbance regimes have degraded ecosystems, quite radically in some cases. Especially in the American West, fire suppression over many decades has resulted in unnaturally intense fires, such as the 1988 Yellowstone fires, resulting in trees being replaced by shrub cover or uncommonly dense stands of fire-tolerant species such as lodgepole pine. In heavily agricultural and developed regions, only small fragments of reasonably intact forest ecosystems can be found, and a fragment of an ecosystem is not sustainable without continual intervention. How forest ecosystems can be restored, even when the trees have been completely removed, is the focus of this chapter.

## The Challenge of Restoring Forested Ecosystems

Forested ecosystems have been degraded, largely, by combinations of five types of stressors:

1. Exploitation. Because forests produce fiber and fuel, they often have been exploited by overcutting, improper harvest techniques, and subsequent mismanagement. On the positive side, good management can protect and restore the ecological integrity of a forest, while providing many benefits, including wood products.
2. Disruption of disturbance regimes. All forests in North America evolved with periodic disturbances from fire, wind, and insect outbreaks, or flooding in the case of bottomland forests. Until rather late in the twentieth century, effort was made to prevent or quickly suppress all forest fires.[2] Ecologists now understand the important role periodic disturbance played in shaping forest communities, and how alteration of these disturbance regimes leads to degradation of forest ecosystems. Historically, fire more often occurred under extreme conditions of low humidity, high wind, and high temperatures. Prescribed fire can rarely be implemented safely under similar conditions. More research is needed to determine the extent to which repeated light fire

or other practices, including timber harvesting, can substitute for natural disturbance.

3. Invasion of nonnative species, including insects and diseases. Globalization has resulted in the introduction of nonnative species to nearly all ecosystems of the world. As technology speeded up travel and commerce, and vulnerability of ecosystems increased because of other stressors, the rate of invasion by nonnative species has increased. In nearly all cases, alien species lack the disease, parasite, and predator controls with which they coevolved in their native ecosystems. As a result, they usually have competitive advantages when introduced into foreign ecosystems. As they spread, they reduce the growth, vigor, and populations of native species, and increase extinction rates.
4. Fragmentation. Forests in North America once were nearly continuous from the Atlantic Coast to the Mississippi, from Panama to the Arctic Circle, and from Nova Scotia to the Pacific Coast. Especially in the Midwest and Southeast, natural forests are now fragmented by highways, cities, agricultural fields, and intensively managed monocultures of pine. In other regions, natural forests have been replaced by plantations of nonnative trees, such as *Eucalyptus* spp. In North America, with the exception of extensive regions in the boreal forests of Canada and Alaska, and relatively small areas in lower 48 states and Central America, most remaining forests, even where well managed, have been cut over several times. A combination of high-grading in the past coupled with selection for commercially valuable species has resulted in younger forests with different mixes of species than were present historically.
5. Herbivory. A wide variety of herbivores, mostly insects, inhabit healthy forests. Introduction of the gypsy moth, among other nonnative herbivores, has had large impacts on deciduous forests in the East. Unnaturally high populations of native white-tailed deer now overbrowse preferred species in virtually all eastern deciduous forests. The high deer population has resulted from fragmented forested landscapes, reduced predation, and urbanization. Additionally, many forests in both the East and West have been heavily grazed by livestock, with ensuing ecological impacts.

Restoration of forests is quite different from restoration of most other kinds of ecosystems. A prairie, for example, can be direct-seeded, and with a decade of care, key structural elements and much native diversity can be restored. Most herbaceous species can go from seed to maturity in one to three years . Most trees require one to several decades, even under ideal conditions, to begin to set seed. Forests have complex structures, usually involving trees of many sizes and ages, and even dead trees in various stages of decay. As trees mature, the environment within the forest changes, with concurrent changes in flora and fauna. This process, even when the trees are planted to jump-start restoration, requires decades. Many forest species cannot be successfully introduced until at least some of the pioneer trees are near maturity. In particular, this includes many of the herbaceous species, epiphytes, and shrubs, not to mention a host of invertebrates and some vertebrate animals. Moreover, the tree species that typically can be established in open sites are different from those that occupy the mature, stable forest. In many cases, only after a hundred or more years can tree species be introduced that will develop with the mature forest. Thus, forest restoration is often a process that requires decades to centuries, depending on the forest and the starting point.

The complexity of the forest ecosystem and the time required for it to reach maturity raises the question of sustainability. Sustainability must always be evaluated within the temporal scale of the ecosystem under consideration. Because forest succession is such a drawn-out process, exceeding human lifetime by two or more generations, it is impossible to conclude with any certainty how sustainable any particular management practice might be, other than practices that move the forest toward the conditions of health found in pristine forests.

## Applying the Principles of Restoration

The first step in restoration of any ecosystem is to take stock of what remains (see chapter 2). This means getting acquainted with the species, the natural communities, the ecological processes essential to the function of that system, and the stressors or conditions that may be negatively impacting its recovery. This is where spending time throughout at least one growing season is essential, walking, observing, and examining the flora, fauna, soils, and water. During this time, it is also important to seek nearby reference areas, on similar soils and topography, to get a sense of less-disturbed forests of the same type.

In analyzing a forest ecosystem prior to restoration treatments, it is important to have a good concept of what a healthy forest looks like. This is all the more important because many people have never seen a healthy forest. Unlike an abstract construct, such as wilderness, *healthy forest* is a functional construct and refers to a forested ecosystem that is able to adjust to natural as well as unusual stresses of disturbance. You might want to review the general characteristics of a healthy ecosystem in chapter 2. They translate into the following characteristics in a healthy forest:

1. High water quality and normal hydrology. If the forest is large enough to contain the headwaters of a stream, the water should be clear, even after a hard storm. In small forested tracts, it is more likely that streams originate off the property, so prior to this assessment, walk the perimeter of the property with a topographic map to get an overview of the landscape. The stream draining a healthy forest will have less variation in discharge rates, although annual variation in discharge will vary with region and watershed.
2. High biodiversity. Each type of forest will have a characteristic composition. A healthy forest will contain a high diversity of native species and few, if any, nonnative species. To evaluate this, you need to compare species lists from reference natural areas of the same forest type in the area. Experienced field botanists in the area may be willing to walk the property with you and probably will have a pretty good idea of what species should be present in that particular landscape setting. While faunal species are also important, they are much more difficult to inventory, so the focus typically is on botanical composition.
3. Diverse, normal structure. As with composition, a healthy forest will contain a structure characteristic of that forest type. At a minimum, this may include bryophytes and ferns; herbaceous species, some of which are ephemeral and found only seasonally or in some years; shrubs; vines; small trees, including understory species as well as regeneration of canopy species; and larger trees. Humid forests will support especially diverse populations of epiphytes, which are plants found growing on other plants (e.g., a resurrection fern on a live oak limb). Many forests will have an overstory of trees taller than the general canopy. Additionally, the structure will include standing dead trees and fallen trees and limbs in various stages of decay, generally referred to as coarse woody debris (cwd). On the forest floor of most healthy forests will be several layers of fine litter. At the top will be the most recently fallen leaves and twigs, with scarcely any decomposition. Beneath will be partially decomposed litter. Nearer the soil will be organic remains of the litter, decomposed beyond recognition. Beneath will be a layer of litter mixed with mineral material. In coniferous forest, the transition to mineral soil will be quite abrupt, whereas in deciduous forests, the mixed organic-mineral layer can be quite thick, grading gradually into lower mineral layers.
4. Diverse communities with good connectivity. A healthy forest must either be sufficiently large to have a range of communities, indicated by variation in composition and structure, or be connected to nearby forests with this added

> dimension of diversity. As climates change, as disease and parasite populations or weather fluctuate, healthy ecosystems must be able to respond. A critical component necessary for healthy response, in addition to the genetic variability contained in the gene pool of each species, is variability in the associations of species and in edaphic conditions—such things as soil, topography, and hydrology. As conditions change, some species will be better suited to survive, some less so. Likewise, some habitats will be more important refugia, some less so. Variation in altitude and aspect or soil texture and chemistry throughout a forest should be reflected in variation in the flora and fauna.

None of the indicators of a healthy forest are black and white, any more than is true of human health. No one can say in absolute terms "I am healthy." Health is a continuum, from very healthy to very unhealthy. The challenge is to evaluate the ecosystem along this health continuum and determine the symptoms or conditions of a healthy forest that are lacking or poorly represented. The next step is to determine why those symptoms are present or what can be done to mitigate or restore the healthy conditions.

Most forests of more than a few acres will have variation in soil and hydrology that is reflected in different relative abundance of species. Always be mindful of variation in both tree species and understory species. Mid- to late spring is often the best time to evaluate these differences. This variation results in different responses to stressors and may require different management, including restoration treatments. Prepare a map of the different types of forest communities, overlaying it on a base map of soils or topography. Note that the ecotone between communities is sometimes abrupt, but more often it is gradual. Where you draw the line is fairly arbitrary when the ecotone is broad. In addition to species differences, you may choose to delineate communities because of different size or age of trees or because of other structural distinctions. List species found in each community type. If possible, extend your community map beyond property boundaries to show the context within the broader landscape.

Once the forest is mapped, each delineated community or unit should be examined and described in detail, and the probable stressor(s) identified and mapped. Continue with goals and objectives, following the steps outlined in chapters 2 and 3.

## Examples Across North America

The USDA Forest Service recognizes twenty-five major forest types in the United States and Puerto Rico (fig. 7.1). In her seminal book, E. Lucy Braun described the

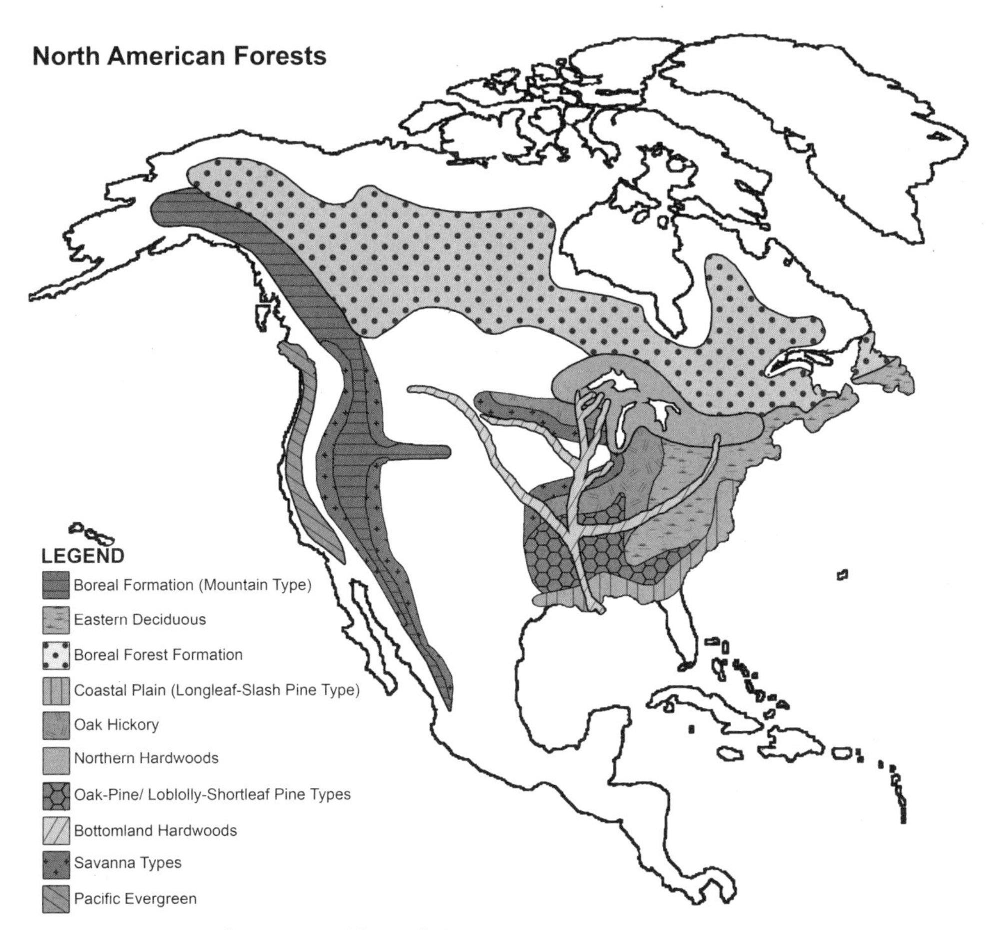


FIGURE 7.1 Major forest types of North America.

original, presettlement forests in the eastern United States.[3] About two-thirds of the Forest Service types are characterized by postdisturbance, early successional species. The other third represent mature, less-disturbed cover types. We will focus on the latter, recognizing the early successional forest types that are found in conjunction with each mature type.

## *Longleaf-Slash Pine Forests*

Longleaf and slash pine historically dominated the coastal plain from central Florida to South Carolina and west to the Mississippi River. The now-rare longleaf-wiregrass ecosystem, once covering an estimated ninety million acres, was maintained by ground fire, with average fire-return intervals of two to four years. Only scattered patches of this remarkably diverse ecosystem remain. In addition to longleaf pine, wiregrass, also called pineland threeawn grass, was a dominant species. Loss of this ecosystem can be attributed to fire suppression, clearing for agriculture and development, aggressive logging of old-growth stands, and conversion to loblolly pine plantations. More than thirty species associated with longleaf-wiregrass communities are threatened or endangered, including the red-cockaded woodpecker. The unusual diversity of the longleaf-wiregrass community is suggested by one study that found 124 plant species in one 10 $m^2$ plot. The U.S. Fish and Wildlife Service, through its Partners for Fish and Wildlife Program, is seeking private cooperators interested in restoring longleaf-wiregrass communities and will pay as much as 50 to 75 percent of the cost.

If remnants of a longleaf-wiregrass community remain, restoration can be as simple as restoring prescribed fire, usually in spring. There is some disagreement on fire frequency, but many studies have shown that the more frequent the fire, the higher the native diversity. Annual burning is not uncommon, although we recommend varying the frequency, from annual to every two or three years. Many plant species remain in the soil seedbank and will reestablish in time. However, because of isolation, small patches usually must be augmented with additional species. Plugs of nursery-grown species can be planted during winter, or seeds of native species from sources as near as possible can be broadcast on recently burned areas. Alien species may need special attention if they are not controlled by prescribed fire.

If restoration must begin with cleared land, the process is more lengthy and challenging. Begin by eliminating all weedy vegetation, usually with broadcast application of suitable herbicides. Note that some weedy vegetation may be woody in nature and require a different herbicide than that which is effective on herbaceous species. Check with the local Natural Resource Conservation Service (NRCS) or U.S. Fish and Wildlife offices for recommendations and requirements. Once weedy vegetation has been eliminated, plant longleaf pine (no more than 250 per acre) with random, nonsystematic spacing, and seed with native herbaceous species from nearby natural areas or commercial sources. Wiregrass, of course, will be a dominant component of the seed mix. Weeds will return, and you may see little evidence of restoration for the first two or three years. Do a light prescribed burn the third year after seeding and then every other year afterward. Continue to augment native species by seeding or planting plugs following prescribed fire. By the tenth year, native species should dominate. As pines mature, thin to maintain an open canopy.

## *Bottomland Hardwood Forests*

Bottomland hardwoods occupy floodplains along lakes and rivers, especially major rivers and streams near the coast. These ecosystems usually flood annually and are important to the health of streams as recipients and filters of floodwater. Soils, as a result, are deep and fertile, although typically poorly drained. Bottomland forests were arguably the most productive forests in North America. Much attention was recently focused on this forest type as a result of reports that the ivory-billed woodpecker, long thought to be extinct, may have been spotted in a large old-growth bottomland forest in Arkansas. Prior to settlement, bottomland hardwoods covered over one hundred million acres, but well over half have been destroyed and the land largely converted to agricultural production. Nearly all have been altered by dams, channelization, and levees that have changed the natural flood regime and/or lowered water tables. Fragmentation of extensive bottomland forests is thought to be the primary threat to ivory-billed woodpeckers and a host of other species that occupy the habitat. Bottomland forests are especially important habitat for many wildlife species, including wood ducks, which utilize acorns and other hard mast that falls into shallow water, through the winter.

A remarkable array of trees and shrubs characterize these forests. In the South, dominant trees include sweetgum; bald cypress; several species of oaks, including swamp chestnut oak, water oak, laurel oak, and overcup oak; water hickory; hackberry; and green ash. In the Northwest, dominants include black cottonwood, Oregon ash, and red and white alder. In the Midwest and Northeast, sycamore, silver maple, and river birch are common. Locally, in the Midwest, bur oak and swamp white oak can occur in open, savanna conditions in floodplains that are subjected to periodic fire (see discussion of savannas later in this chapter).

As a result of disruption of the flooding regime that was critical to the development and sustainability of floodplain forests, alien species commonly invade them, and shifts occur in relative dominance of native species. For example, in the Midwest, native prickly ash and nonnative glossy buckthorn can form thickets that are nearly impenetrable. Garlic mustard is a serious problem in the East and Midwest, especially where the flood regime has been greatly altered. Garlic mustard gets a quick start in the spring when soil would ordinarily be under water, and once established, seeds are quite persistent. Reed canary grass can develop near-monoculture stands in the absence of annual flooding. Trees are much less successful regenerating in these conditions. It is not clear if the annual flooding would have prevented buckthorn, but it did deter prickly ash, garlic mustard, and reed canary grass. In many cases, fire also occurred in bottomland forests during late summer or autumn. Fire provides good control for prickly ash and is a deterrent to garlic mustard and buckthorn until they become so well established that they act as deterrents to fire.

Ideally, restoration of bottomland forests will include restoration of the flood regimes associated with them. Because that involves restoration of the watershed and removal of flow-control devices (see chapter 9) such as dams and levees, it usually is not possible. Moreover, development in most watersheds results in more rapid runoff and a change in discharge rates that would cause different flooding regimes, even without flow-control devices. While it is probable that in the very long term, perhaps a century or more, soils will change as a result of the altered flooding regime, such changes are so slow as to be undetectable over a few decades. Nevertheless, restoration usually must be approached without opportunity to restore historic flood regimes.

In some instances, high water can be trapped behind levees to simulate annual flooding. Ditches and drainage systems can be disabled to create wetland conditions more similar to presettlement conditions. To some extent, prescribed fire can be used when flooding is no longer possible but with careful monitoring to be sure the effects will shift the ecosystem in a favorable direction. Along the Wisconsin River, we have found that prescribed fire keeps prickly ash and buckthorn under control and maintains the open understory in which many tree species can regenerate. Herbicide may be necessary to initiate control, after which fire can help maintain favorable conditions. The preferred herbicide is glyphosate because it is nontoxic and quickly breaks down in the soil. A 3 percent solution of the active ingredient is sufficient to control most herbaceous species when applied as a foliar application during the growing season. A backpack sprayer works well for clumping species such as garlic mustard. For more extensive stands such as reed canary grass, especially if some native species may be hidden beneath, consider a wick application. A fibrous wand saturated with herbicide is swept across the unwanted vegetation, avoiding contact with desirable vegetation. For woody species such as buckthorn, avoid treatment in the early growing season when plants are least susceptible; late summer or autumn is preferred. Apply an 18 percent solution of glyphosate to freshly cut stumps using a brush or small sprayer.

If a good mix of tree and shrub species characteristic of bottomland forests are present, attention can be focused entirely on alien species and maintaining native diversity, including a healthy balance of tree species. It may be possible to use periodic timber sales to help maintain a good balance of species. When bottomland forests must be established on cleared land, it is essential to address hydrology first. Usually, land would not be cleared if high water were a frequent limitation. While it may not be possible to restore annual flood regimes, it is necessary to restore annual high water, sufficient to at least saturate the soil surface. Failing this, it will be futile to try and establish a bottomland ecosystem. One could easily establish mixed stands of early successional species, such as sycamore, sweetgum, or river birch, and these may diversify in time, but they likely will become heavily invested with alien species and never regain the mix of species that characterized the native bottomland forest community.

If the hydrology can be reasonably repaired, sites should be treated with herbicide to remove alien species, then planted with a diverse mix of woody species characteristic of local bottomland forests. Many species can be direct-seeded. Attempt to match the diversity found in local remnants. Frequent monitoring will be required to catch early establishment of alien species, with use of herbicides to eradicate them. Experimentation will be necessary to see what hydrologic conditions, fire, herbicide, and mechanical treatment combinations will be most favorable for moving the ecosystem toward diverse stability.

## *Oak and Loblolly-Shortleaf Pine Forests*

In a broad band of the coastal plain and Piedmont Plateau, from Virginia to east Texas, forests were dominated by a mix of oaks and loblolly pine. Following disturbance, loblolly and shortleaf pine usually become dominant. On abandoned agricultural fields, relatively pure stands of shortleaf pine are common for the first fifty to one hundred years, with oaks and associated hardwoods gradually increasing. Much of the area was once cleared for cotton production. Many former agricultural fields have been planted with genetically improved loblolly plantations grown intensively on rotations as short as twenty-one years.

Although characterized by oaks of several species, many other tree species coexisted in the oak-pine forests. Of the oaks, white oak, southern red oak, post oak, and blackjack oak were most common. Mockernut and pignut hickories and American holly were common associates. American chestnut was important along the northern edge of the oak-pine type prior to its demise by the chestnut blight, an introduced fungus disease.

The stressors in the oak-pine type are primarily exploitation and overgrazing. These forests were typically high-graded, commonly having been cut over three or more times in the last century. It was common for livestock to be turned into these forests, including swine that were fattened in the autumn on the hard mast. The stress associated with livestock in forests results more from compaction and consequent root disease than from consumption of vegetation. Swine, however, also are disruptive because of rooting and disturbance of surface litter and soil. Because soils of this region tend to be high in clay, they are more vulnerable to compaction. Compaction leads to lower infiltration and greater runoff, with associated erosion, and loss of soil aeration. Most trees that grow in well-drained soils are sensitive to soil compaction. Effects of soil compaction can persist for decades after livestock have been removed. Moreover, high populations of deer now make regeneration of oaks difficult in some areas. Where forests were cleared for agriculture, years of tillage and intensive tobacco, cotton, and soybean farming led to extensive erosion on these clay-rich soils, and organic matter is typically quite low. Full restoration cannot be achieved until the soils have largely recovered, but this happens most rapidly under forest conditions.

Fire history of this forest type is different from other types. Historically, ground fire had a return interval of twenty to fifty years, with hotter fires once or twice a century. It was the periodic hot fires that maintained loblolly and shortleaf pine during presettlement time. Fire increased with settlement, used to clear land for agriculture and to favor forage for cattle. As more people moved into the region, spring burning became a cultural tradition. People often did not have a specific reason to burn, or burned to "control snakes" or "keep the woods open." The more frequent fire regime altered ecology, favoring oak and pine, and disfavoring others such as red maple and tulip poplar. Frequent fire, coupled with livestock, further increased erosion and loss of topsoil.

Restoration of oak-pine forests involves removing livestock, restoring a fire regime more typical of presettlement, and harvesting trees to favor the distribution of species more similar to that which was present before settlement. In time, topsoil will be restored and soil compaction lessened. Careful research will be needed to determine which species should be favored and which not, in each instance. Sources such as early survey records, reference natural areas, and descriptions by the earliest naturalists in the region will provide some clues. Although some species will have been extirpated, fragmentation is much less a problem with this forest type than with some others. Invasive species ordinarily are not a great problem unless there has been extensive disturbance. On abandoned agricultural land, however, weedy species, including nonnatives, may be well established. Honeysuckle, kudzu, and black locust can be locally abundant and require aggressive control.

Species such as honeysuckle and kudzu can be controlled with foliar application of glyphosate (as discussed earlier) during the growing season. Black locust, on the other hand, requires application of 18 percent active solution to fresh-cut stumps. All stems in a clone, which is sometimes quite extensive, must be cut and treated. Otherwise, root sprouting will occur, and resulting stems will require follow-up treatment.

If starting with oldfields, abandoned agricultural land, the easiest route of restoration is through loblolly and shortleaf pine plantations. In some cases, natural stands of shortleaf pine may develop, or be encouraged by direct seeding into disked fields. As the canopy closes, weedy plants will be eliminated. Unless the field is quite isolated, oaks and associated species will slowly become established. Planting seeds of these species may speed the process, although squirrels are remarkably adept at finding them. Pines should be thinned to release hardwoods, and this may involve a commercial timber sale under some conditions. Periodic light fire, beginning in ten to fifteen years, will favor native species, and reduce the risk of severe fire. Full recovery of the ecosystem will likely require one to two centuries.

## *Oak-Hickory Forests*

Oak-hickory forests are the most common type in the eastern deciduous forests. As the name suggests, this type was dominated by several species of oaks and pignut hickory. The more common oaks were white, black, northern red, scarlet, and chestnut. This was the forest type where American chestnut was most important. Its decline resulted in increase in other species, especially chestnut oak. In addition to pignut hickory, mockernut occurs in the southern half, and shagbark throughout much of the area. Other important species included tulip-poplar, sourwood, and redbud.

Oak-hickory forests are ecologically similar to the oak-pine type. They were maintained by periodic fire, with historic return intervals of twenty to fifty years. Hot fires also occurred, probably less frequently than in the oak-pine. Recruitment of cohorts of oaks and hickories often occurred following moderate fire. When seed sources are present, pitch and shortleaf pine often regenerate. Following severe disturbance, pioneer tree species such as sassafras, tulip-poplar, staghorn sumac, eastern red cedar, and shortleaf and Virginia pine become established and are slowly invaded by trees characteristic of mature forests.

Stressors in this forest type are exploitation in the form of clear-cutting and conversion to alternative uses such as agriculture and development, as well as repeated high-grading for the most valuable timber trees. Grazing of livestock and frequent burning also were common in many areas. These forests are still adjusting to the loss of American chestnut, once a dominant species. Gypsy moth, introduced near Boston from Europe in 1869, has spread through New England and into the oak-hickory forests. Oaks are especially attractive to gypsy moth larvae, and repeated defoliation weakens trees, making them more vulnerable to drought and disease. In some areas, disturbances such as logging have greatly increased oak wilt, a fungus disease that is lethal to members of the red oak subgenus.

Strategy for restoration of oak-hickory forests is similar to oak-pine. Long-term protection from livestock, careful forest management, and selection harvesting are required. Common buckthorn and garlic mustard are increasing rapidly in the oak-hickory region and should be controlled aggressively. Also, black locust, ailanthus, kudzu, and honeysuckle vine are local problems. (See previous discussion for treatment.)

One of the challenges is overabundance of red maple, associated with increased fire suppression. Especially in the more northern part of the oak-hickory region, lack of fire also favors the more shade-tolerant sugar maple. Red maple readily stump-sprouts when cut, and sprouts tend to grow more rapidly than those of oaks or hickory. It also has greater shade tolerance and is a prolific seed producer. Sugar maple is one of the most shade-tolerant deciduous species, and in the absence of any disturbance, will increase when a seed source is present. In the southern Appalachian region, eastern white pine can also become dominant at the expense of oaks and hickories. Prescribed fire, coupled with careful harvesting and precommercial thinning of undesirable trees may be required to maintain the balance of species desired.

Restoration of an oak-hickory ecosystem from abandoned agricultural land will require at least a century. Restoration can be jump-started with a plantation of Virginia or shortleaf pine, or by eliminating weeds and direct seeding tulip-poplar. Establishing tulip-poplar is probably a quicker way to get back mixed hardwoods, but may require mulching to get a good stand established. Alternatively, we recommend letting oldfields revert naturally with gradual increase in sassafras and sumac. Both sassafras and sumac can be increased by planting seeds or root fragments. Attractive clones will soon follow. Plant acorns, hickory nuts, black walnuts, and persimmon seeds to encourage more diversity and hasten forest development, and scatter tulip-poplar seed in cleared patches. Where nearby stands exist, eastern red cedar may also become established in oldfields. Spot-treat weedy infestations, especially kudzu and honeysuckle vine. Native goldenrods, asters, and broomsedge will slow the establishment of forest species but provide a beautiful mixed woody-herbaceous community for decades. Following this avenue for restoration is an excellent example of being patient and working with nature rather than trying to overpower natural processes.

## *Northern Hardwood Forests*

Northern hardwood forests, as discussed here, include forests with dominant components of sugar and red maple, American beech, birch, basswood, and white ash. Depending on the area and local conditions, other important components may include eastern hemlock, eastern white pine, northern red oak, white oak, and hophornbeam. With severe disturbance, aspen can dominate the forest for seventy-five to one hundred years. Intensive forestry in the region focuses largely on aspen using coppice regeneration from root sprouts, with clear-cutting cycles of forty to fifty years. Northern hardwood forests occupy the northern tier of states from New England to the Great Plains, south of the boreal forests.

Fire was much less frequent in northern hardwood forests than in forests to the south or west. Winter snow, followed by wet spring conditions discouraged fire except for brief periods in the autumn after leaf fall. Although infrequent, seldom occurring more than once or twice in a thousand years, severe windthrow was ecologically important, especially in creating opportunity for establishment of extensive stands of eastern white pine. Trees such as white pine are long lived, and once established, they can persist for three or more centuries. Natural succession primarily involves recruitment of tolerant species, such as sugar maple and hemlock, or gap-phase species such as basswood that can grow rapidly through small holes in the canopy created by local windthrow or disease.

The ecology of this forest type changed dramatically with extensive forest harvesting through the region. Attempts to establish agriculture were marginally successful, and forest land cleared for agriculture often was later abandoned. Especially in the upper Midwest, much of the abandoned land came under ownership of county, state, and federal agencies, which reestablished red pine plantations or allowed natural succession to restore forests. Much of this land continues to be managed for short-rotation aspen or pine. Most the forests continue to be routinely harvested, and aspen, the mainstay of the paper industry, has become the most common cover type. As a result of intensive forestry, white-tailed deer populations have reached unnaturally high densities, and species of both woody and herbaceous plants favored by deer have been reduced.[4]

Even where forests have been severely disturbed by intensive harvesting, they generally will recover without much management. Old-growth aspen of seventy-five or more years is soon replaced by mixtures of tree species, depending on site and seed source. Because this region has been less fragmented, most species still remain, at least in refugia, often not far from any site. Management, therefore, can focus on controlling deer and reintroducing of species missing from a site.

If starting with an agricultural field, restoration can be either passive, by accepting aspen that usually will become established, or active, by planting conifers. Aspen will be slower invading oldfields with heavy grass cover. Herbicide may be necessary to prepare these sites for trees. Aspen becomes the nurse crop for other forest species that will become established over the next one to two hundred years. The alternative approach involves planting white or red pine. Red pine creates a less conducive environment for establishment of forest species than white pine. Mixtures, including white spruce, can also be used. Direct establishment of a hardwood such as northern red oak is difficult without weed and deer control, and mixing hardwoods with conifers is generally not very successful. Plantations of conifers will control weedy populations, and over time, permit the establishment of other forest species. Conifers, unlike aspen, can be thinned and sold to gradually open the site for other species. Aspen are usually not thinned because of damage to residual stems. If aspen are clear-cut, they will reestablish from root sprouts and the process of restoration will start over. If the aim is a more diversified northern hardwood forest, and seed sources for native species are lacking, underplanting aspens will be necessary to begin the conversion from the aspen forest to one that is more diverse.

## *Boreal Forests*

The boreal forest is circumpolar, occupying a band in North America across southern Canada into Alaska, with extensions down the Appalachian, Rocky Mountain, and Cascade ranges. Species tend to vary in the southern alpine extensions, although the ecology is much the same. Mature forests are dominated by spruce and true firs, while early-successional forests are dominated by quaking aspen, and jack and lodgepole pine. A variation of the boreal forest occurs in the northern Great Lakes region, with eastern white pine and red pine mixed with spruce and fir. Except in western alpine areas, paper birch is usually present to varying degrees in both early and late successional forests.

The boreal forest is fire maintained, with historic fire-return intervals of fifty to one hundred years in uplands. Old-growth conditions begin developing in the absence of fire after a hundred and fifty years. As forests age, spruce and fir increase, along with periodic mortality, especially from spruce budworm. Accumulation of coarse woody debris, ingrowth of young spruce and fir, and periodic mortality greatly increase susceptibility of the boreal forest to fire. It is rare for boreal forests to survive as long as four hundred years without fire. Wind is another, although much less frequent, disturbance. Whereas fire favors fire-adapted, early-successional species such as jack and lodgepole pine, wind tends to release advanced spruce and fir regeneration, and paper birch, which is more wind resistant. Disturbance in forests with even a small component of aspen will result in prolific regeneration of aspen, primarily from root sprouting.

Major stressors on the boreal forest are climate change and exploitation. However, because the boreal forest is so extensive and occupies a more northern, less-populated region, extensive harvesting has not threatened this ecosystem except in local areas. Large amounts of petroleum and natural gas reserves occur under boreal forests, and continued cutting and fragmentation is inevitable. Already, there are huge areas that have been clear-cut. If aspen is present, clear-cuts tend to redevelop quickly but primarily with aspen rather than with the conifer species, as would be the case following fire. If aspen is not common, clear-cuts will remain quite open, with sedges, asters, beaked hazel, and alder providing most of the vegetative cover. Global climate change is likely to stress the boreal forest more than most other forests because conditions are changing more rapidly and to a greater extent at higher latitudes and altitudes. Warming conditions will increase exposure to fire, favoring more early-successional species. Drier conditions will put greater stress on spruce and fir, increasing budworm outbreaks, which will lead to more and hotter fires. Fire favors jack and lodgepole pine.

In much of the West, alpine boreal forests are extensively grazed. Grazing has two primary effects. It keeps the forests more open and limits fine fuels, thereby reducing fire frequency. Second, grazing selectively reduces populations of more palatable herbaceous species. The alpine forest type most affected is probably lodgepole pine, an early-successional species that is maintained by stand-replacing fires. In rare instances of long-term absence of fire, Engelmann spruce and subalpine fir begin to invade lodgepole communities, perhaps encouraged by bark beetles (family: Scolytidae), which are encouraged by often extensive stands of lodgepole, especially when trees become stressed.

Boreal forests cannot be restored except on a large scale. Periodic crown fires are essential and require broad landscape approaches. Where cutting has favored aspen, forests can be allowed to return through succession to spruce-fir. Aspen will begin to die at age seventy-five to a hundred years, with gradual shift to spruce and fir dominance. If the first generation of spruce-fir is lightly stocked as a result of inadequate seed source, the next fire cycle is likely to restore a more natural balance of species. Recently burned areas should be planted or seeded with jack or lodgepole pine, if these species are not represented in the immediate area. In the Great Lakes region, eastern white pine and red pine can be included in the mix.

Where clear-cutting has resulted in brush-dominated landscapes because aspen was not adequately represented, planting with jack or lodgepole pine may be necessary. If areas are heavily vegetated with hazel, alder, raspberries, or grass, site preparation with herbicide may be required. Alien species are generally not a severe problem in boreal forests, although we have seen local areas badly infested with common hemp nettle. Such areas may require spot treatment with herbicide.

## Savannas

Savannas are open forests, usually with less than 50 percent canopy cover, with diverse grass and forb groundcover. Seen from a different perspective, savannas are grasslands with scattered trees. They have some characteristics common to both forests and grasslands. In a broad zone between the eastern deciduous forest and midwestern grasslands, oak-dominated savannas occupied millions of acres during presettlement time. Fragments of oak barrens or savannas also occurred from New England throughout the oak-hickory region, primarily on shallow or coarse soils, especially on south-facing slopes. Oak savannas were once very widespread in California. Dominant tree species, depending on the site and region, include black oak, Hill's oak, white oak, and valley oak. Important associated species include sassafras and sumacs. On heavier soils in the Midwest, bur oak was often dominant. Bur oak and swamp white oak also occurred in floodplain savannas where fire was a frequent occurrence. These sites were more droughty and fire prone. Ponderosa pine forests along the Rocky Mountain chain and pinyon-juniper forests of the Southwest are other expressions of savanna. The latter is characterized by pinyon pine and several species of junipers. This woodland or savanna type alone once occupied about thirty million acres, including most of the Colorado Plateau.

Evidence is clear that savannas were maintained by frequent fire, and over hundreds of years savannas ebbed and flowed with shifts in climate. During hotter and drier periods, savannas shifted toward the cooler and wetter side of the broad ecotone in which they occurred; during cooler and wetter periods, trees invaded and savanna species survived on the warmer and drier side.

Savannas were easier for settlers to develop than forests. Fewer trees had to be removed to facilitate agriculture. Grazing was there for the taking. Many savannas, however, occurred on such coarse or shallow soils that they were not suitable for agriculture. These were called *barrens* in the East and Midwest.[5] Similarly, poor soils in the pinyon-juniper forests discouraged tillage, although they have been extensively grazed by livestock.

With settlement, savannas became increasingly fragmented and isolated. Fire suppression became more effective. With fire suppression, ingrowth of small trees and shrubs occurred, including species that were less fire tolerant. In time, once-open savannas that escaped development in the East and Midwest became closed forest fragments within farmland. As this occurred, savannas became less fire prone and more susceptible to nonnative species. Especially in the East, common and glossy buckthorn invaded. Ironically, grazing helped protect savannas against invasion of trees and shrubs, and many of the best examples remain because of livestock grazing. Some have speculated that bison may have had a historic role in maintaining savanna structure and composition.

Ponderosa pine savannas have had a parallel but different history. Because they occur in the drier regions of the western mountains, from southeastern British Columbia to Arizona, they have been less affected by development. Like oak savannas, fire-return intervals of two to twenty-five years kept them open.[6] Recruitment of trees was probably a chance occurrence, although in oak savannas evidence indicates that cohorts of new trees were periodically recruited, probably during periods of wetter climate when fires were less frequent. As livestock were increasingly introduced to ponderosa savannas, the light fuels that carried frequent ground fires were reduced, and fire frequency declined. Moreover, fire suppression became a priority in management. As a consequence, more young ponderosa survived and increased the ladder fuels leading to severe crown fires. As this was occurring, cheatgrass was introduced to the West and rapidly spread across the range of ponderosa pine. Cheatgrass provides decent forage for a brief time early in the growing season, but soon dries and becomes unpalatable. The cheatgrass, however, increased the light fuel between patches of ponderosa that were becoming more vulnerable to fire. At the same time, cheatgrass reduced the native diversity of mountain meadows and exposed slopes to increased erosion.

Pinyon-juniper woodlands and savannas vary from closed woodlands, especially at altitudes approaching 7,000 feet where pinyon pine dominates, to very open savannas dominated by junipers at 5,000 feet. As with all savannas, they were especially important to native people as sources of food and fiber. Harvest of the edible seeds of the pinyon pine was cause for annual celebration. Other than harvesting for fuelwood, many now view pinyon-juniper negatively because it encroaches on rangeland. Ironically, heavy grazing favors the junipers that livestock find unpalatable, and woody vegetation has increased steadily since heavy grazing on the Colorado Plateau began over a hundred years ago. Draconian measures such as chaining, the pulling of heavy chains stretched between two bulldozers, have been used to reduce woody cover.

Restoration of oak savannas, including barrens, has been successful with a combination of structural manipulation and fire. Opening the tree canopy as quickly as possible is key. This can be done by a selection thinning of smaller trees, especially species less adapted to savannas, sometimes as a commercial timber sale. Or, it can be done by running a very hot fire through an overgrown savanna. The latter is more risky but probably closer to the role historic fire had in these forests. In either case, follow-up with repeated fire is essential. Once the canopy is opened to about 50 percent cover, fires can be fairly light, but are needed every two to four years in most cases to control woody growth and encourage fire-adapted grasses and forbs. On coarser soils, the seedbank will usually be adequate, but on heavier soils, many species may need to be reintroduced.[7] Frequent fire may be all that is necessary to control alien species, but if buckthorn is well established, it will have to be hand cut and stumps treated with 18 percent glyphosate or other suitable herbicide.

In restoring ponderosa forests, timber sales might also be feasible if access to the site and markets is suitable. Using timber sales to remove ingrowth in roadless areas, however, can be done only by pushing roads into previously isolated areas. The expense typically is offset by clear-cutting, or at least cutting the mature trees, which are the ones restoration would preserve. Consequently, building roads into remote ponderosa forests may be economically feasible but not be helpful in restoration. Use of prescribed fire is much more difficult in the West, as well, because of more variable wind and large tracts of highly flammable fuels. On a small scale, ponderosa can be hand thinned to open them, with follow-up prescribed fire under cool, damp conditions. On a larger scale, it may be best to await the inevitable hot fire that will often completely open the forest. Planting ponderosa pine early in the following growing season can jump-start the recovery, or if sufficient natural regeneration occurs, thinning the young forest early followed by careful use of prescribed fire under cool conditions can work in some locations.[8] Also, prescribed fire can be applied in early spring, following retreat of snowmelt up mountain slopes, where snow will act as a fire barrier to limit spread into alpine forests. Once ponderosa forests are established, prescribed fire is needed every five to ten years to maintain the open savanna condition.

Control of cheatgrass has been essentially impossible on any scale. Fire tends to favor it. Herbicides that would eliminate cheatgrass will also kill native species. On a small scale, glyphosate can be used to kill cheatgrass and any other herbaceous species growing with it, followed by introduction of desirable species through planting or seeding. However, this is expensive, time consuming, and likely to fail because of nearby cheatgrass seed sources. It is likely that the ecology of western forests and meadows is forever changed as a result of cheatgrass, unless some biological control is found. This handicaps the use of prescribed fire since cheatgrass is a flash fuel that will quickly carry fire from one area to another.

Restoration of pinyon-juniper also is primarily a matter of waiting for wildfire. Pinyon-juniper savannas are now often seen as woodlands, but they will burn sooner or later, and when they do, there is an opportunity to begin management. Once burned, they can be maintained by prescribed fire, especially if grazing is controlled. Monitoring of grazing also is necessary to ensure that the most palatable species are not being eliminated. With controlled grazing and prescribed fire, this ecosystem can recover, although it is likely that many species have been locally extirpated because of widespread grazing. More study is needed to evaluate the ecological changes in pinyon-juniper savannas and suggest recovery methods.

## *Pacific Evergreen Forests*

From central British Columbia to northern California, evergreen forests dominate the natural landscape along the coast, extending into the Sierra Nevada Mountains, where coastal vegetation gives way to alpine forests. The northern part of this forest complex is sometimes referred to as *temperate rain forest* with annual precipitation between 150 and 360 cm (60 and 140 inches). Pacific evergreen forests drew much attention with the controversy surrounding the endangered northern spotted owl that was federally listed in 1990. Douglas-fir is the most common tree species. It is a dominant or codominant in several forest types of the region where it is associated with Sitka spruce, several species of true firs, western red cedar, incense cedar, and western hemlock, among dozens of others. Toward the southern end of the Pacific evergreen forests, oaks and other hardwoods that are more fire tolerant become increasingly common, and Douglas-fir becomes less important.

Douglas-fir is an early-successional species, although it can live for several hundred years and reach immense size. It has no equal in timber productivity in North America. Without disturbance, it eventually is replaced by more tolerant conifer species. Thus, these forests are commonly referred to as *mixed evergreen forests*. Diversity is remarkably high, and species vary greatly from the north, where precipitation is high and evaporation low, to the south, where precipitation becomes more seasonal and evaporation is extreme during the dry season. Fire frequency and intensity vary along this climatic gradient.

Historically, succession was most commonly initiated by fire, ranging in return intervals of twenty to four hundred or more years, depending on climate. Fire frequency increased from north to south. Frequent fires are more restricted to fine fuels near the ground, whereas less-frequent fires are more often stand-replacing crown fires that favor regeneration of Douglas-fir. Douglas-fir is most common in the midrange of Pacific evergreen forests, where stand-replacing fire frequency approaches the age of mature Douglas-fir forests.

Few old-growth stands now remain, and it is this that triggered the controversy over the spotted owl, which was once thought to be largely restricted to old-growth conifer forests. The timber industry, which favors the extraordinarily productive Douglas-fir, now replants most harvested areas, and plans on forest rotations of approximately one hundred fifty years.

Restoration of Pacific evergreen forests is problematic because of the difficulty of using prescribed fire. As a result of urban sprawl and fragmentation, and the very high value of timber, fire suppression is especially aggressive. As in the boreal forest, it is difficult to manage these evergreen forests on a small scale using fire. On the positive side, invasive species are not a severe problem, and dominant trees can potentially live for hundreds of years. Thus, it is possible to protect areas, even without fire, and expect that they will be relatively stable for many centuries. From north, where fire may be largely unnecessary, to south, prescribed fire becomes increasingly important. In California, especially, light fires every few decades are necessary to encourage and maintain optimal diversity and discourage alien species.

Connectivity is also important, particularly in the more fragmented areas that have been heavily developed. Larger tracts, or areas linked to other forested areas, even if they are managed for timber, will help maintain diversity.

## Conclusions

Although over 300 million forested acres have been cleared for agriculture and development, forests still cover approximately a third of the United States. Nearly twice as much forest cover occurs in Canada, although most is boreal forest. Old-growth forests and forests other than boreal forests in truly good health are rare in North America as well as in much of the rest of the world, in part because of exploitation but also because of air pollution and invasive species.[9] The complex nature of forest structure and composition makes it possible to regularly harvest many types of forests and still protect their ecological integrity. Stressors such as altered disturbance regimes and exploitation lead to loss of species and different structure with a resulting loss of integrity. With loss of integrity, forests lose some ability to protect watersheds, sequester carbon, provide critical habitat for some fauna, and often even maintain aesthetic value. Restoration, commonly thought of as ecosystem management within the forestry community, can assist in the repair of damage to forested ecosystems, even when the entire forest community has been removed. However, it takes decades to centuries to restore ecological integrity, depending on the forest and degree of damage it has suffered. While the same steps and principles apply to forest restoration as to any other type of ecosystem restoration, the time frame to achieve measurable success in a forest is much longer.

# Chapter 8

# Wetland Restoration

*Poets who know no better rhapsodize about the peace of nature, but a well-populated marsh is a cacophony.*

Bern Keating

Few ecosystems offer more restoration rewards than wetlands. They are among the more important habitat types for harboring diversity, as well as for protecting surface-water quality. Wetlands support a disproportionate percentage of the rarest plants and animals on earth. Consequently, they have received attention from conservationists, ecologists, and lawmakers alike.[1] Historic descriptions of the biological richness and abundance of life of wetlands are nearly universal.[2]

Even the smallest wetland can attract a pair of ducks and shorebirds, not to mention salamanders, frogs, and interesting insects. Great blue herons will kite down to hunt frogs in the edges of wetlands no bigger than your living room. Raccoons will slink along the edges, absently feeling with their feet for crayfish or other edible morsels. Wetlands are important habitats for amphibians and many insects and are used by a host of species that split time between aquatic and terrestrial habitats. Less appreciated is the role of wetlands in buffering storm surges, and releasing runoff water slowly into streams. Floodplain wetlands are especially important in mitigating flood surges.

Wetlands are broadly distributed in the world, being integral parts of nearly every terrestrial biome (fig. 8.1). They vary from cold-water seeps that form fens, to riparian zones along streams, to black-water swamps in the South, to blanket peatlands that occupy extensive landscapes in cold climates.[3] Wetlands are important habitats in arid regions, such as the basins in southern Wyoming and around the Great Salt Lake in Utah.

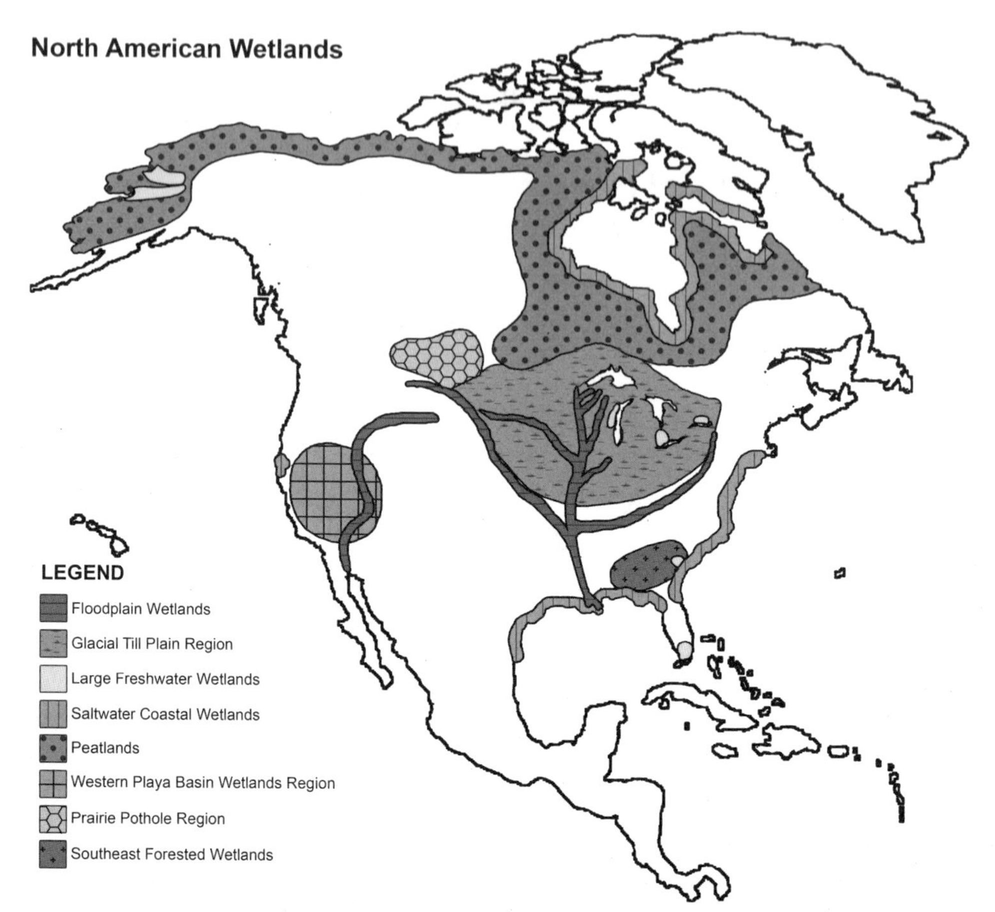


FIGURE 8.1 Wetland regions of North America.

Filling and draining wetlands have been long-standing public concerns. Millions of acres of wetlands in the Midwest (United States) have been drained by tiles and ditches to claim the land for agriculture. Extensive flats that supported wetlands in the southeastern United States have been replaced by loblolly pine plantations. Wetlands that hosted thousands of species now support a single plant species: wheat around the prairie potholes of the Dakotas, corn or soybeans from Indiana to Nebraska, loblolly pine in Florida and Mississippi, and fruits and vegetables in California.

Because so many wetlands have been drained, most of us have little appreciation for how much of the historic landscape was wetland. For example, an early account tells of a person being able to paddle a canoe in the spring from central Illinois to Chicago. Now, traveling that same route by car, one sees primarily corn and soybean fields. Thousands of miles of drainage ditches and tiles have contributed to incalculable loss of habitat and biodiversity.[4]

## Wetland Form and Function

Wetlands are a critical part of the earth's hydrologic system, receiving water from snowmelt and rain, slowly releasing it from the land to recharge streams and lakes. Some wetlands, in permeable substrates, also help restore groundwater. However, most wetlands have little or no groundwater recharge function. Fine sediments and layers of partially decomposed organic matter found in many wetlands create seasonal or perennial seals that limit infiltration. Additionally, wetlands occupy landscape positions at or below water table.

Wetlands are very dynamic. Some have no standing water, only saturated root zones. Others are inundated with emergent vegetation, such as trees or cattails, and submerged aquatic vegetation. During dry years ephemeral wetlands may support terrestrial species, often annuals that germinate sequentially in bands around a central pool as water levels drop.[5] Riparian wetlands are particularly dynamic. When a stream floods, the water spills into the wetland, river velocity decreases and suspended sediments are dropped. In relatively undisturbed watersheds, streams carry little sediment, and deposition is minimal, in contrast to the heavier sediment loads carried from agricultural and urban landscapes. Plant species in floodplain wetlands are adapted to tolerate the alternation of saturated and unsaturated soil and periodic deposition of sediments. The filtering of surface water, now considered an important function of wetlands, was much less important in most historic watersheds when erosion was far less than now.

Wetlands can also be found around springs and seeps, where groundwater keeps the soil saturated for much of the year. These wetlands are usually in gently sloping topography, often in glacial till. Various species, including many rare plants, especially sedges and orchids, are delicately adapted to the temperature and chemistry of the emerging groundwater.

The water level in wetlands usually is related to precipitation patterns. During dry periods, wetlands can dry sufficiently to allow fire to sweep across them, reducing shrub or tree cover. In other years, the same wetland may stand deep underwater. Some remain wet continually, while others are wet for only portions of each year. Wetlands usually lose diversity when water fluctuation is altered from their natural hydrologic regimes. Even a static water level, caused by outlet control structures, such as culverts, can result in loss of diversity.

The predictable daily inundation cycles of water level in tidal marshes results in some of the most productive ecosystems on Earth, in part because they are bathed regularly with nutrient-rich water. While plants are not as diverse, being dominated primarily by two species of salt-marsh grass, they form the seaward edge of one of the most dramatic ecotones. Following creeks inland, the water becomes increasingly less brackish and often in a short distance upstream the ecosystem will shift from coastal salt marsh to brackish and then freshwater marsh or forests. This ecotone represents important habitat for an extraordinary number of species and is a productive ecosystem for fish and shellfish, including species such as shrimp that support vital economies.

Mangrove forests are another important type of coastal wetland. Over fifty mangrove species occur worldwide, although Florida has but three. Red mangrove forms thick tangles along the seaward side of these wetlands, where the salt concentration in the water is greatest. Black mangroves occur on slightly higher ground, where salt concentrations are less. White mangrove is farthest removed from seawater, in least brackish areas. Together, these trees stabilize shorelines against storm erosion, and mangrove roots create a unique habitat where the young of many aquatic species find refuge.

## Planning Wetland Restoration

The cleaner the water entering the wetland, the higher quality the wetland will be. A diverse, stable wetland requires that water be cleaned before it enters the wetland. If the watershed is healthy, the wetland is usually healthy. To protect and maintain a quality wetland, the larger tributary landscape must often be addressed. When the ecology of a watershed is damaged, wetlands will reflect those changes. This is easily observed around many metropolitan areas. Wetlands that are surrounded by developments quickly degrade into cattail or reed canary grass or giant reed grass monocultures.

Vegetation of a wetland reflects the hydrology and the chemistry of the water, and therefore the watershed that feeds it.[6] Indeed, pollen from cores taken from wetland substrates, and the thickness and nature of individual layers that are laid down each year, often extending through tens of thousands of years of accumulation, have been used to interpret the vegetation, fire frequency, and extent of historic disturbance in a watershed. Saturated organic matter decomposes very slowly and can accumulate over centuries. On the other hand, peat or muck exposed when a wetland dries can sometimes burn for months or years. If enough organic sediment burns, the substrate surface is lowered, a pond develops, and the process of accumulation begins anew. Nutrients released by the fire can stimulate the germination of seedbanks of some wetland species.[7] Historically, many wetlands have undergone repeated cycles of wetting, filling, drying, and burning through long climatic cycles.

## Regulations and Legal Definitions

In the United States, all wetlands now are protected by the federal Clean Water Act (1972). In some states, counties, parishes, townships, or municipalities, there are secondary regulations that also protect or regulate the use of wetlands. Before undertaking wetland restoration, understand the laws and regulations that may limit what you do or how you do it. You should also read chapter 9, because stream restoration is governed by some of the same state and federal regulations as well.

Given their dynamic nature it is not always apparent where a wetland exists or where its edges or limits are. There is no alternative to field investigation to delineate a wetland. You may have sufficient familiarity with soils and vegetation to proceed on your own, but if not, request assistance from your county conservationist at the local NRCS office, or hire an experienced consultant to do the delineation for you. Because wetlands are highly regulated, permits likely will be needed for doing restoration work, and they require formal delineation.

You should also be aware that while agricultural and forestry lands are generally exempt from federal Clean Water Act requirements, most states also regulate dredging, filling, or draining of wetlands. Because of the overlapping regulations and multiagency jurisdictions, wetland restoration must start with understanding the regulatory controls that apply to the project.

There are many definitions of wetland, including academic and technical definitions that differ in both minor and important ways. Three examples follow:

> Clean Water Act, as defined by the U.S. Army Corps of Engineers: Those areas that are inundated or saturated by surface or ground water at a frequency and duration sufficient to support, and that under normal circumstances do support, a prevalence of vegetation typically adapted for life in saturated soil conditions. Wetlands generally include swamps, marshes, bogs, and similar areas.[8]

> U.S. Fish and Wildlife Service (FWS): Wetlands are lands transitional between terrestrial and aquatic systems where the water table is usually at or near the surface or the land is covered by shallow water. For purposes of this classification, wetlands must have one or more of the following attributes: (1) At least periodically, the land supports predominately hydrophytes; (2) the substrate is predominately undrained hydric soil; and (3) the substrate is nonsoil and is saturated with water or covered by shallow water at some time during the growing season of the year.[9]

> Textbook: Wetlands are lands on which water covers the soil or is present either at or near the surface of the soil or within the root zone, all year or for varying periods of time during the year, including during the growing season. The recurrent or prolonged presence of water (hydrology) at or near the soil surface is the dominant factor determining the nature of soil development and the types of plant and animal communities living in the soil and on its surface. Wetlands can be identified by the presence of those plants (hydrophytes) that are adapted to life in the soils that form under flooded or saturated conditions (hydric soils) characteristic of wetlands.[10]

It is the definition in the Clean Water Act that provides the guidelines for federal regulations. In 1987, this was expanded to include the three characteristics noted in the FWS definition. To qualify as a wetland by the earlier federal standards, all three criteria must be present. The FWS definition is much broader, stating that only one of the criteria must apply. The textbook definition also indicates that both hydrophytes and hydric soils characterize a wetland. Because of their dynamics, wetlands challenge classification, which has resulted in the difficulty in regulating and protecting them. Creating technically accurate and legally defensible wetland classifications has been a challenge, resulting in numerous lawsuits and many revisions of policies.

The following is a simplified classification key pointing up some important differences among wetland types based on substrate and vegetation:

## Wetland Types

A. Forested
  1. Mineral substrate
     (a) Seasonally inundated: primarily, flatwoods and floodplain forests.
     (b) More or less permanently inundated: swamps such as Okefenokee, portions of the Everglades, or red mangroves where currents or deposition prevent buildup of organic layers.
  2. Organic substrate
     (a) Seasonally inundated: ephemeral wetlands in protected watersheds with little erosion.
     (b) More or less permanently inundated: e.g., pososins, tamarack, and spruce bogs, black and white mangroves swamps, etc.

B. Nonforested
  1. Mineral substrate
     (a) Seasonally inundated: shrub carrs and sedge meadows, including floodplain meadows that regularly receive mineral deposition.

   (b) More or less permanently inundated: shrub carrs and sedge meadows that receive regular mineral deposition.
2. Organic substrate
   (a) Seasonally inundated: portions of tundra, seeps, alpine meadows, and sedge meadows along streams in watersheds with little disturbance.
   (b) More or less permanently inundated: peat bogs, coastal wetlands, and depressions that rarely receive mineral deposition.

As is evident from the classification key, the primary differences among wetlands result from hydrology, water chemistry, and landscape context and conditions, including degree and extent of upland erosion. These differences, in turn, result in different plant and animal communities. Depth and persistence of water strongly influence biodiversity of vegetation; macroinvertebrates; and breeding birds, including waterfowl.[11]

## Substrates

Inorganic substrates such as clay, marl, and silt tend to be associated with the most productive and nutrient-rich wetlands. These wetlands also are often in areas more likely to receive nutrient enrichment as a result of agricultural activities.

Wetlands that have been drained by agricultural tiles usually have organic-rich mineral substrates suitable for tillage after they were dewatered. Often, restoration simply involves eliminating or disabling the tiles. These wetlands are typically low in nutrients and tend to have high-quality plant communities, though not necessarily the highest diversity.

Organic substrates include the accumulated peat that may have recognizable plant parts and fibers remaining (fibric peats), or they may have decomposed to a muck with few recognizable plant fibers (hemic peats), or they may have no recognizable plant tissue (sapric peats). In nearly all cases, the pH can vary from neutral but trend toward acidic to strongly acidic, and most nutrients will be unavailable. It is useful to know the type of substrate in a wetland being restored. It tells a lot about the history of the wetland and the watershed and also portends the likely future conditions.

## Water Quality

There are five general categories of water quality, with unlimited variations among them:

1. Cool, mineral-rich groundwater, most common in fens and some springs. The pH is often neutral or slightly alkaline.
2. Cool, mineral-poor groundwater, most common in seeps or springs. The pH varies, but most commonly is slightly acidic.
3. Warm, nutrient-rich surface water, common in floodwater, or water draining from urban or agricultural areas.
4. Warm, nutrient-poor surface water, more common in floodwater from relatively undisturbed watersheds.
5. Nutrient-rich salt water, with tidal fluctuations and storm surges leading to frequent inundation.

The types of minerals and concentrations greatly affect the kinds of plants and how well they grow, as well as macroinvertebrates such as snails and dragonflies. It is the mineral extremes, the richest and poorest, that often are associated with the rarest species and the most unusual wetland communities. Wetlands fed by water that is neither nutrient rich or nutrient poor support communities that are more ubiquitous.

Warm, nutrient-rich water is often very low in dissolved oxygen, and this greatly affects the macroinvertebrate population. Wetlands fed by warm, nutrient-rich water tend to develop low-diversity communities, often dominated by one or two species, such as cattails in deeper water or reed canary grass in shallow wetlands or those seasonally wet.

Groundwater may be depleted of oxygen and is colder during the growing season than surface water. Emerging water can give rise to a wetland that serves as a refugium for heat-sensitive plants and animals. If the flow continues year-round, the wetland also may provide protection from harsh winter conditions in colder climates.

## Hydrology

In addition to water source and quality, water-level dynamics strongly influence the type and nature of the wetland. A simple classification using hydrology criteria is as follows:

### *Classification of Wetlands*

A. Permanently inundated or saturated to the surface
   1. Freshwater
      (a) Acidic, as a result of slow and incomplete decomposition: e.g., bogs and pososins.

(b) Near neutral or alkaline: e.g., fens and seeps with water from limestone or till.

2. Brackish or saltwater
    (a) Tidal
        (i) Inundated twice daily by high tides: Low salt marsh, dominated by *Spartina alternifolia*, and mangrove swamps.
        (ii) Inundated only by storm tides, and salt water commonly diluted by freshwater: High salt marsh, dominated by *S. patens*.
    (b) Inland saline wetlands, e.g., extensive basins in Nevada, Utah, and California, and even salt water upwelling in freshwater wetlands (e.g., Montezuma National Wildlife Refuge, New York).

B. Seasonally inundated or saturated to the surface
    1. Ephemeral ponds and wetlands
    2. Tundra, both high latitude and high altitude
    3. Ephemeral seeps
    4. Floodplains, including sedge meadows, shrub carrs, and forests
    5. Flatwoods, especially coastal

Sustained water for longer periods (durations), with periodic dry-down, will favor submerged aquatic plants and trees such as some mangroves, bald cypress, or water tupelo. These trees tolerate seasonal inundation but germinate and become established in saturated, but exposed, soils. Some herbaceous plants can spread by underground roots or rhizomes even beneath water, but few species can get established in statically deep water, or successfully spread into the deeper water from margins. Cattails and reed canary or giant reed grass are notorious for colonizing shallow impoundments. Wetlands that always or usually have saturated soils but are rarely inundated, such as fens and seeps, support some of the most diverse communities.

Some wetlands have critically important cycles of hydraulic dynamics. The regeneration of most trees and shrubs in swamp forests occurs during infrequent droughts, perhaps only one dry period in thirty to a hundred years or more. The regeneration of marshes that have been dammed by beaver, where cattails have become prevalent, will cycle and regenerate plant diversity when the dam fails. Sedge meadows and shrub carrs will often burn during infrequent dry periods, perhaps only once a decade. The infrequent fires maintain these communities that otherwise would eventually succeed to forest cover.

There are predictable natural patterns in water-level dynamics in most wetlands, which are easily disrupted in altered watersheds. A curve depicting the change in water level over time is called a hydrograph. Hydrographs can be developed for wetlands, streams, or lakes. For an artificial lake, the hydrograph measures the quantity

**Annual hydrographs and normal average water levels for restored wetlands**

Engineered vs. Natural System

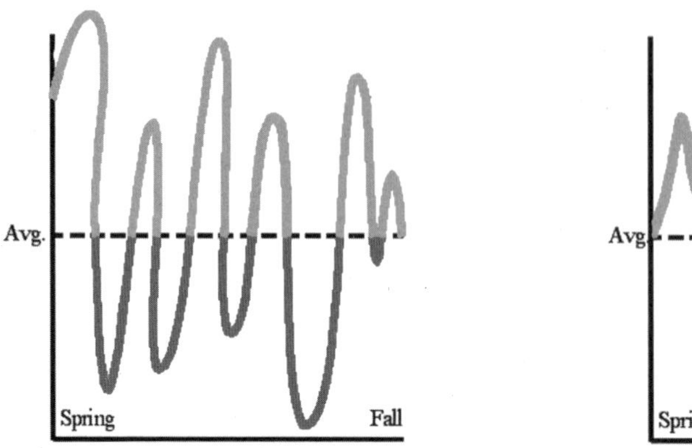

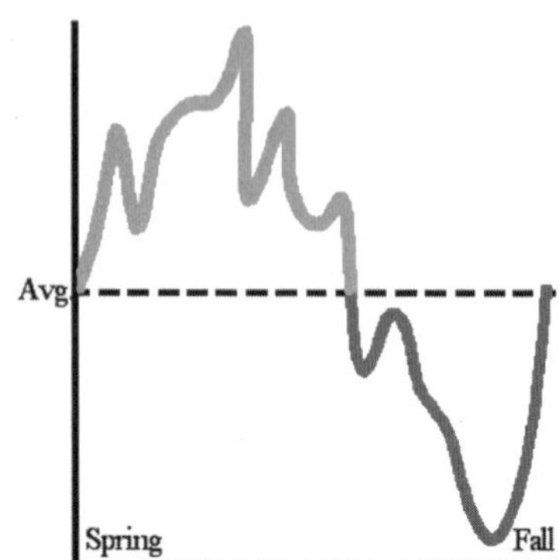


**Engineered System**
*Creates unpredictable swings in water levels
*Creates biological instability
*Promotes habitats for invasive species
*Promotes poor water quality

**Natural System**
*Creates annual seasonal high and low
*Creates predictable hydraulics and seasonal trajectory
*Promotes habitats for a stable yet diverse range of native plant and animal communities

FIGURE 8.2 Stylized hydrographs comparing controlled wetlands, rivers, or lakes in disturbed watersheds with those of largely unmodified watersheds.

of water flowing over the spillway. Figure 8.2 shows stylized hydrographs for a wetland in an altered watershed compared to one in a stable, largely unaltered, watershed. Note the unpredictable water level changes, called *bounce,* in wetlands located in disturbed watersheds. The hydrograph of a largely unaltered watershed, called *nature's hydrograph,* shows water-level changes associated with the winter snowmelt and spring rains that typically combine to create predictable seasonal high- and low-water levels. These wetlands have a higher diversity of plants and animals. In altered wetlands, invasive plants and animals are more common, and muddy, unstable shorelines are often a telltale sign of water levels that quickly change or bounce. The management and maintenance needs in restored wetlands with different types of hydrographs are quite different. Wetlands exhibiting nature's hydrograph are low maintenance in contrast to wetlands with unpredictable hydrographs of a disturbed watershed.

The amount of bounce in restored midwestern (United States) wetlands generally should not exceed two feet more than a normal seasonal high-water level. Where bounce regularly exceeds two feet in those wetlands, there is a consistent decline in plant and animal diversity.

## Physical Gradients and Biological Change

Both topographic and chemical gradients affect plant and animal diversity. Water depth determines the type of plant community present. If water depth drops off steeply, wetland zones will be narrowly aligned against shores. If there is seepage water that is different chemically or thermally, this can give rise to different plant and animal communities. For example, where mineral-rich cold water seeps into warm nutrient-enriched surface water, a gradient forms where the two water bodies mingle. Localized dilution gradients, even if noticeable springs are not present, are common. These can influence vegetation margins around wetlands, for example, a sedge zone.

Gradients are not static. Consider saltwater gradients along coasts, as described earlier. In northern climates especially, springs often create thermal gradients including areas that remain unfrozen. Often, ice freezes into the underlying substrates, then during the spring thaw, it "plucks" the substrates and plant materials frozen to the underside of the floating ice. This creates pileups of floated material on the downwind side of an open-water wetland. The locations from which material is plucked become deeper, and these microdepressional features become home to species of plants and animals favored by deeper water.

Wildfires once regularly burned across much of North America, and where they encountered wetlands, fire burned as far as possible into the underlying substrates. Deeper water moats on the upwind side (usually west and southwest) of many midwestern wetlands were a result of prairie fires burning into the wetlands, and substrate sampling in these moats shows high levels of ash, carbonized plant remains (including charred wood fragments) and higher nutrients. There are other patterns in wetlands that reflect current and past changes. Effects of animal activity is seen in trails created by deer, beaver canals, beaver and muskrat lodges and deeper water around them, dead trees from raised water levels, and wet meadows created when wetlands drained after beaver dams were abandoned.

## Stressors

We use the same step-by-step restoration process for wetlands as described in part 1. This begins with the exploration of the wetland to be restored. Initially, this is done to identify the *stressors* that are currently affecting, or that historically may have affected, the wetland and associated watershed.

Many of the common stressors in wetlands are interrelated. As a result, even a minor change in one stressor can result in shifts in others. Except for alteration of hydrology, such as drainage of a wetland or an increase in runoff water being directed into the wetland, effects are often subtle, with shifts in species composition. Changes in hydrology, especially a result of runoff from agriculture and development that is frequently associated with increased deposition of sediments and macronutrients, will commonly cause native diversity to decrease, and increase the risk of invasives.

Depending on the stressor and the magnitude of the change, a wetland may not show substantive change for many years. However, if runoff from a feedlot were to be introduced to a high-quality wetland, for example, even without a significant change in hydrology, the wetland would degrade rapidly. The most common changes that create stress for wetlands and that are key to understanding and implementing wetland restoration are the following:

1. Drainage. Dewatering by ditches, tiles, or other alterations of hydrology, or diverting water away from a wetland.
2. Diversion of water to wetlands. This commonly is associated with development or changing land use in the watershed, often with associated change in erosion and nutrient runoff. Even when erosion in a watershed is not increased, accelerated runoff can alter the hydrology in a wetland.
3. Water-control changes, such as from damming, or road construction and culverts. Also included here is static control and loss of natural wetting and drying cycles.
4. Excessive nutrients and sediments. This is usually associated with agriculture or urban development in the watershed. It most often is interacted with altered hydrology.
5. Invasive and introduced species. Common examples include purple loosestrife, common reed, and glossy buckthorn.
6. Decrease in groundwater sources. Often associated with increased surface water entering the wetland. As wells deplete groundwater, increased runoff may maintain more or less the same amount of water entering a wetland, but the hydrology and chemistry will be different.
7. Salinization. While salinity increases naturally in confined wetlands over time in drier climates, this process can be greatly speeded up because of runoff and changes in the watershed of a wetland. Also, salt applied to roads as a deicer often ends up in adjacent wetlands.

## Restoration Strategies and Process

Wetland restoration requires determining what stressors can realistically be reversed or mitigated without deleterious effects on neighboring properties. For example, one would not want to disable a drain tile only to learn that a neighbor relies on the same tile for his farm field. As is also the case with river restoration, the restoration of wetlands requires attention to conditions in the watershed, and often, negotiations among neighboring landowners and regulators before, during, and after restoration work begins.

The restoration process for wetlands follows the same program laid out in chapter 3, with some added detail and attention to the physical drivers summarized in preceding sections of this chapter. To emphasize where the additional attention is needed, consider the following points:

- Understand changes in hydrology and their effects. The relationships between hydrology, topography, and substrates become very important in wetland restoration design. Typically, a topographic map of the potential restoration location is completed. This includes careful mapping of the soil boundaries; property lines; and locations and elevations of tile, ditches, and other drainage infrastructure. Borings are often necessary to reveal underlying substrates, especially where excavation is proposed to create or recreate a wetland basin. The map should indicate locations of springs or indicators of the location of historic springs (e.g., marl or elevated muck deposits); locations of buildings, roads and their elevations, septic systems, and other buried utilities. By restoring wetlands, we often make the land wetter and therefore change the buoyancy of buried utilities such as high-pressure gas lines. This can be a real threat, as lines can break. Make sure that the restoration strategy will not result in higher flood elevations that might back water onto roads or rail lines, or into nearby septic fields.
- Investigate seedbanks and sources of biotic diversity. Many plant seeds remain long lived in the substrates of a wetland. You may be able to complete a restoration without adding seed or reintroducing plant species. In bigger wetland restorations, do surface and subsurface (to about 30 cm depth) substrate collections, place the samples in flats in a greenhouse and grow out, identify, and enumerate the native and invasive species from the soil samples. This helps determine what species are needed, what will come up on their own from the seedbank, and what invasive species present a management threat in the project. If you sample substrates with georeferencing (e.g., use global positioning system [GPS] to map the locations of the samples), you can use response data to map the likely vegetation in the restoration. This can help determine the invasive species management needs and locations, and also where you may want to add seed for species that will not appear from the seedbank.
- Obtain needed permits. The early coordination and preparation of the

documentation necessary for any permits is another task peculiar to wetland restoration. Make sure you explore these needs during early meetings with the county conservationist, U.S. Army Corps of Engineers, Department of Natural Resources, or other county or state regulatory agents.

- Clean water ensures higher quality wetlands. This cannot be overstated. If surface water feeding the wetland compromises nutrients and sediment, it becomes a very high priority to develop effective pretreatment before allowing the water to enter the wetland. A simple method is to create an area in the landscape with managed vegetation, such as perennial grass, that serves as a sediment trap or biofilter. Because most landowners do not control the entire watershed feeding a wetland, this is a common necessity. It must be sized to effectively pretreat the water entering the wetland. Sometimes, this is all that is required to restore a wetland.

If large portions of the watershed are in forest or perennial grass (ideally, native prairie or wetland cover), then chances are good that the quality of the water entering your wetland will be acceptable. Understanding the condition of the water may require mapping the watershed, or at least the portion through which water passes in route to your wetland. Map the watershed as accurately as possible, focusing on the types of land uses present. Use this process to ascertain the likely quality of water the wetland will receive. It is also advisable to look carefully at the drainage routes to see if there are sediment deposits. Visit the watershed during a major storm event or during spring runoff to witness the quality of the water flowing in the drainage courses. Is it turbid? Are there large quantities of suspended sediments or debris? If so, expand the restoration project to consider restoration of critical upland buffers, perhaps with prairie, forest, or small wetlands. These biofilters expand the overall diversity of the restoration and ensure a higher quality wetland project.

## Conclusions

Wetland restorations may seem daunting because of the technical challenges of restoring hydrology and dealing with the regulations that are commonly required. Professional assistance may be needed to help delineate the wetland, analyze the hydrology, and ensure that drainage on neighboring properties will not be negatively altered. Once basic hydrology conditions are reestablished, however, wetlands usually respond rapidly.

Determining feasibility is essential in the initial design of a wetland restoration. Deciding if the hydrology can be restored, and to what degree, is often a challenge.

Commonly, problems originate in the watershed beyond property ownership, necessitating cooperation with neighbors or intervention in runoff. There are many tradeoffs in design, construction, and management of wetland restorations. For example, if you live where water rights cannot be reallocated adequately for complete restoration, a seasonal wetland to take advantage of early pregrowing season water may be the only realistic goal. If you live where water is plentiful, you must ensure that the topography will support the restoration on your property without inadvertently causing flooding or backing up septic fields on neighboring property.

Wetland restorations can transform a landscape within several weeks or months to a place with abundant wildlife. Acre for acre, few habitats have greater diversity or contribute more to the diversity of an overall landscape. Thus, while the challenges are many, the rewards of wetland restoration justify considerable effort.

# Chapter 9

# Stream Restoration

*In the confrontation between stream and the rock, the stream always wins—not through strength but by perseverance.*

H. Jackson Brown

Streams are a barometer of the health of their watersheds. The story of streams, as with so many natural resources, has been one of exploitation and lack of understanding. Few streams throughout the world have escaped pollution, channel modifications, and increased flooding as a result of mismanagement or development in the watershed. Fortunately, as with any ecosystem, streams can be restored if the stressors in the watershed can be mitigated.

Streams were a focus for human development long before the Industrial Revolution. As natural avenues for commerce, rivers represented the superhighways of the nineteenth and twentieth centuries, but before that, villages nearly always developed along streams, which facilitated transportation and provided energy, food, and water for domestic use. As increasing technology spurred industry and urban development, streams were increasingly seen as cheap ways to deal with waste. Thus began a radical divergence in the functionality of biogeochemical cycles of natural ecosystems and those of human-dominated ecosystems.[1]

## Underlying Problems

Waterborne disease, as much as anything, began to call attention to the exploitation of streams. The eutrophic conditions Steve found in the stream at Stone Prairie Farm are a common characteristic of nutrient-enriched streams in agricultural and urban landscapes. Nutrients come from runoff of fields where fertilizers have been applied, from feedlots, and from septic or sewage treatment facilities. Many industries release organic waste in their effluents, further stressing the aquatic ecosystem. Stimulated by high nitrogen and phosphorous in the water, algae and other aquatic plants flourish, greatly exceeding the consumption by snails and other animals that feed on aquatic vegetation. As water warms during the summer, it holds less oxygen, but respiration and decomposition speed up. Decomposition switches from aerobic to anaerobic processes. Most species of fish and macroinvertebrates cannot survive under these anoxic conditions. Restoration requires that the organic waste and nutrient loading be eliminated or greatly reduced.

Dams, originally built to generate hydropower, became an engineering solution to flood control. Channelization and levees were intended to move water more quickly out of a watershed to minimize local flood damage. Sedimentation increasingly occurs behind dams as watersheds are cleared of natural vegetation that previously absorbed and retained water. With this process, hundred-year floods become seventy-five-year then fifty-year floods, meaning they occur with greater frequency. In addition to the flooding problems that result from development, rapid climate change will also result in altered flow regimes, increasing flooding in some watersheds and reducing water entering many others.

Stream restoration requires that both the stream and the watershed be restored. Too often, only stream-bank erosion or flood control is addressed. Excessive erosion and floods are symptoms of a watershed that is not functioning properly. Solutions such as rip-rap, channelization, and levees are only Band-aids. Stream restoration always works best within the context of planning and design of the watershed. Recall the principle of working at the proper scale on ecosystem processes. With most watersheds, this involves working with many landowners. Landowners in the headwaters may see little economic benefit of watershed restoration, while downstream property is being damaged by floods and poor water quality. The resulting disparate benefits of restoration must be balanced by incentives to the upstream landowners to modify their land-use or waste-management practices. Government incentives often lead to lengthy negotiations, reviews, and planning that can take years.

Aside from ownership challenges, invasive species are an increasing problem. Because stream corridors cross landscapes, they are common avenues for the spread of both invasive terrestrial and aquatic plants and animals. Many nonnative species found along the banks or in the channel invaded through rivers. Salt cedar is an example. This woody shrub now colonizes the banks along thousands of miles of streams in the western United States. Collectively, this species transpires millions of gallons of water that would otherwise be available to support aquatic life or use for irrigation. River restoration commonly must include invasive species management.

Most streams were historically fed by both groundwater and surface water from precipitation, including snowmelt. In some streams, groundwater may have contributed over 90 percent of the water that flowed through the stream during an average year. In developed watersheds, water that once percolated into the soil now quickly runs from the land to the neighboring streams, lakes, and wetlands, often increasing the frequency and severity of floods. Floodwater carries more soil with resulting increases in erosion and sedimentation.

Faster runoff from the land means greater water volume and higher velocity in stream channels. When water velocity doubles, the amount of eroded substrates it can carry increases thirty-two-fold, and the size of rock that can be moved increases sixty-four-fold. Thus, increased velocity equates to greater erosiveness, on the land as well as in the channel and along stream banks. Catastrophic reshaping of a stream channel occurs during peak flood events that may happen only once or twice a century in some climates, but the work of maintaining and structuring the form and pattern in a channel is associated with frequent storm events, primarily those occurring regularly, every one to two years.

The challenge in stream restoration is to restore the capacity of the watershed to hold and more slowly release water. The process for restoring a watershed is covered in the other chapters of part 2. This chapter focuses on the stream channel. Stream restoration, unlike restoration of most other ecosystem types, usually requires coordination across ownership boundaries. The owners as well as other stakeholders will have different interests and potential benefits from the restoration. Technical information and data are needed to properly determine tasks with the optimal cost-benefit ratio. Considerable knowledge may be needed to convince regulators to grant permits, and people skills are necessary to convince those with different interests to work together. Because streams are, by nature, extremely dynamic, good data on which to base arguments are difficult to obtain. Nevertheless, the fact that successful stream restoration can be done is attested to by successes throughout the United States. The Kickapoo watershed in southwest Wisconsin was one of the first. Started in the 1930s, the rolling farmland in the watershed was the focus of the first regional soil conservation effort. Many soil conservation techniques, such as sod-waterways, contour strip-cropping, and terracing, along with crop rotation to build up and maintain soil organic matter, were first implemented there. Farms were revitalized, and the Kickapoo Valley became an attractive recreation area. Although the Kickapoo still floods, water quality has improved immensely, and floods are much less severe.

## Some Principles

Restoration should begin by determining critical changes that have occurred in the watershed and the responses of the stream. This can be done with historic aerial photographs and review of hydrologic data. In some cases, modeling of watershed hydrology can suggest cause-and-effect relationships between the stream and the watershed. As with any restoration process, you begin with field assessments. You may need hydrologic measurements to allow better evaluation and comparison with reference streams.

Research on stream hydrology has established clear relationships between the quantity and rate of water discharge and the shape, dimensions, and dynamics of the channel.[2] These physical relationships are consistent in rivers throughout the world. Resulting predictive models can be used to evaluate the restoration potential, feasibility of various strategies, and likely response of channels as a result of restoration. Several useful evaluation and design procedures have resulted from pioneering research by Luna Leopold.[3] Leopold's work provided a foundation for understanding stream behavior, summarized here:

1. Streams can be viewed as machines that are doing the work of moving water and sediments across landscapes under the influence of gravity. In essence, streams are conveyors. A dynamically stable stream represents the most efficient way this machine can dissipate the energy contained in the moving water. When it is disrupted, it seeks to regain its efficiency in dissipating energy. For example, if you add material to the channel, the stream will move it. If you straighten the river, it immediately begins to create new meanders.
2. The shape, depth, and capacity of streams to move water and sediments are continuously refined by the work of the stream, manifested in erosion and deposition. Strong universal relationships exist among the channel width and depth, the watershed area, and the characteristics of discharge of water from the land.
3. If you overload a river with water or with suspended load, just as with a conveyor belt, problems develop. For rivers, this means shifting of the channel (where not constrained by bedrock), and movement of more material downriver to be deposited in lower pools or floodplains. An example of the energy dispersal resulting from overloading a stream can be observed by laying a garden hose in meanders on a level sidewalk. With a gentle flow of water through the hose, it remains stable and unmoving. If a high volume of water is introduced to the hose, excess energy is released by wild flailing from side

to side. This is analogous to a more rapid shifting of the channel when the capacity of a stream is exceeded. This happens when water is released in higher volume and greater velocity from a watershed that has reduced capacity to retain it. More rapid release of water also increases the suspended load carried in the water.

Unlike hoses, healthy rivers have floodplains, which can receive and hold the additional volume of water in the stream and receive some of the suspended load, which then is dropped as the velocity of water decreases when it spills onto the floodplain. Floodplains are a relief valve for water in excess of what the channel can hold. Levees isolate rivers from their floodplains, increase the scouring and erosion along the channel, and result in the flood surge being passed more quickly downstream where it often causes even greater flood-related problems.

All streams are dynamic, but the rate and extent of change depends on climatic patterns and the conditions in the watershed. As natural ecosystems develop, their capacity to retain water, nutrients, and minerals increases. Human activities and use of resources in a watershed reduce this capacity. Even if climate were to remain constant, human activities in a watershed result in more runoff and erosion.

## The Process of Stream Restoration

Stream restoration requires more data, more paperwork, and more negotiation than most other kinds of restoration projects. Most states require permits for even the simplest component, such as bank stabilization. Large projects must have cooperation from state and local regulators. The U.S. Army Corps of Engineers, the U.S. Department of Agriculture (USDA), and state departments of natural resources or conservation are nearly always involved, along with any affected municipalities.

Restoration often requires the identification and measurement of *reference reaches*, stable sections of the stream (e.g., equivalent of *reference natural areas* in other chapters). Some measurements require training in collection and calculation of data. For more information on field measurements and doing the follow-up computations, refer to Stream Restoration Design National Engineering Handbook.[4]

While the process and restoration principles are fundamentally the same as any other restoration process, because of the regulatory requirements, there are differences. While one might design the most technically correct restoration solution, it may never happen unless the political and regulatory issues are covered. In some cases the technical and regulatory requirements of doing stream restoration might be avoided. Landowners living in rural America should discuss their concerns with their local conservationist or with stream restoration experts. However, be wary of the advice you receive. It often is directed toward symptoms, rather than the underlying causes.

Unless your interest is only in stabilizing a short stretch of stream bank, or similar small-scale project, full-scale restorations should follow the process described here. Depending on the scale and complexity of the project, you may be able to skip some steps. Regulators will typically give you their version of this process and what they require for a permit to be issued for any restoration project. There is more measurement required for streams than with other kinds of ecological restoration. Reference reaches are important to understanding the form and dimensions of stable sections of streams, which are characterized with a range of physical measurements.

Phase I. Problem definition and data gathering

1. Define the problem.
2. Assess conditions through measurement and classification.
3. Assess changes and dynamics through air photo interpretation, upstream and downstream.
4. Define a stable condition by finding and measuring reference reaches.
5. Evaluate what can be accomplished on the property without addressing watershed-wide problems.
6. Meet with agencies to begin discussions on regulatory needs.

Phase II. Measurements, modeling, planning, and design

7. Assess stream channel measurements and characterizations on site.
8. Assess sediment and substrate characterizations on site.
9. Engage in hydrology measures or modeling.
10. Map unstable and dynamic zones, deteriorated vegetation, and erosion risks.
11. Develop working drawings or digital models.
12. Develop preliminary design options for stabilization and restoration of the channel and floodplain.

Phase III. Testing, evaluation, and permitting dialogue

13. Develop preliminary plans for restoration, including plan form, cross sections, longitudinal profiles, erosion control, grading, planting plans, and monitoring programs to document channel stability.
14. Evaluate by modeling sheer stresses and stability in the proposed channel modifications and the role of revegetation in stabilizing channel and banks.
15. Meet for continued discussions of regulations.

Phase IV. Final conceptual plan preparation and permitting submittals

16. Prepare preliminary engineering plan as scaled drawings.
17. Prepare permit packages for state and federal agencies.
18. Obtain permits.

Phase V. Installation and monitoring

19. Finalize construction plans and specifications.
20. Contract for services.
21. Post bonds, if required.
22. Implement restoration earthwork, erosion control, and native plantings, as needed.
24. Install and implement monitoring program.

The challenge is to determine the least invasive restoration strategy to create and stabilize stream habitats. There are many books that address various aspects of stream restoration, but most are focused on in-stream structures aimed to improve fish habitat. These tend to be costly, can be temporary, and often overlook the primary problems. The most viable techniques that successfully encourage a native fishery usually are focused on restoration of riffles, pools, and runs. We do not encourage development of artificial structures. For example, where erosion has resulted in loss of overhanging bank habitat, wooden and metal structures are sometimes installed. We have encountered remnants of these artificial structures while canoeing: metal fence posts or rebar dangerously present in the channel, where they collect debris and often lead to further bank erosion over time. It is better to assist damaged riffles and pools by use of rock or wooden debris, such as fallen logs, repositioned in the stream to create favorable habitat. Sometimes removal or restriction of cattle is all that is necessary. We have had good success with bioengineering, planting live dormant stems of wetland shrubs, such as willow, to stabilize banks in lieu of rip-rap or rock-covered banks that eventually will fail. Finding ways to encourage native plant species enables the stream to restore the missing habitats. Look for ways to work with nature. Always bear in mind that the stream is a reflection of the entire watershed, and its problems as well as the solutions usually lie upstream.

River restoration is more complex than restoration of smaller streams, but both benefit from thinking and attempting to work at the scale of the watershed. While this often is impossible because of the many landowners involved, and political and regulatory jurisdictions crossed by the stream or river, it is worth keeping this overall strategy in mind. Sometimes mentioning the desire to work at the larger scale, to address what is happening upstream, can be enough to inspire neighbors to participate in restoration.

## Conclusions

Throughout human history, we have dealt with streams with contempt or ignorance. Floodplains became fertile land for our intensive agriculture. The forests provided wood for our taking. We put barriers across streams to create enough head to turn our gristmills, sawmills, and turbines. When streams flooded, we confined them by walls, or straightened channels to move water away more quickly. We circumvented rapids with locks so we could motor up and down rivers more easily. We used the water to cool our machines and as a cheap and easy way to dispose of waste.

In the relatively short life span of humans, it is easy to overlook the multimillennia life span of streams. Streams move mountains and cut valleys into which major cities can easily fit. With this perspective, we should expect that human attempts to control streams will be futile and fraught with problems. As with all restoration, we must understand the natural forces and processes that drive the systems, and work with them.

Restoration must assist the stream to regain its physical balance after restoration of the biological communities in the watershed. The nature of land tenure and watersheds usually requires that stream restoration involves many property owners with different benefits and costs associated with the stream. Very often, stream restoration means overcoming mythology and engineering dogma of how to tame the unruly behavior of the stream. It is imperative to work effectively with the human community, businesses, industries, and landowners, all of whom have an interest in the watershed, but each with his or her own perspective. Stream restoration projects can be great unifiers, bringing people together through the watershed linkages. These linkages are much more organic than the typical commercial linkages in human commerce, and more like the linkages among species within a natural community. Much has changed since the pioneers settled along the shores of a river because of the fertile soil, beauty, power, water supply, and transportation. Then, the river was vital to the commercial interests and lifestyle of the community. After a century of deterioration of the watershed and abuse of the river, communities are beginning to realize that the river is a measure of the way we choose to live, a barometer of our quality of life.

Especially in urban areas, people have become increasingly isolated from the streams that were the magnet that drew early pioneers to those locations. In the United States, this divorce began a century or more ago, and now most urban areas are culturally and physically separated from the resources once provided by the river. Restoration must restore a healthy relationship between people and the land.

## Chapter 10

# Desert Restoration

*To say nothing is out here is incorrect; to say the desert is stingy with everything except space and light, stone and earth is closer to the truth.*

William Least Heat Moon

Deserts are perhaps the most misunderstood of all ecosystems, although they occupy about a fifth of the land surface of Earth and have expanded worldwide over the past century. Many consider deserts to be desolate. Consistent with that misguided notion, both the noun and verb are derived from the same Latin root, *deserere*, to abandon. While some arid regions of the world appear to be nearly devoid of life, most deserts, including all in North America, are far from barren.

Many desert areas in North America once supported much greater diversity. Even as early as 1926, Shelford reported a lack of high-quality desert remnants, noting damage from overgrazing and invasive species.[1] Overgrazing, primarily by cattle and sheep, led to a decrease particularly in grasses, while less palatable plant species increased. Many of the plants that have increased are protected by spines or laced with noxious chemicals.

In part because of changes in and around deserts, and because information about their historic conditions is often lacking, biogeographers and ecologists commonly disagree about presettlement characteristics and even the distribution of each type of North American desert. This is especially true of the so-called cold desert in the Great Basin of Utah, Nevada, and Wyoming. The historic ecology of this landscape, now dominated by giant sagebrush, remains unresolved.

## What Is a Desert?

The technical definition for desert is not particularly revealing: "a landscape that receives less then 10 inches (250 mm) of unevenly distributed precipitation annually."[2] Deserts occur worldwide, from the Arctic to the Tropics. Nevertheless, they are commonly classified only as either cold or hot. Technically, much of the Arctic is desert, but because of very low evaporation and extremely short growing season it is generally viewed as a separate biome. Confusion about desert classification also results from the overlap in distributions of many plant and animal species. Effects of topography on precipitation patterns, especially rain shadow effects of mountain ranges, temperature and wind effects on evaporation, and soil variation lead to complex climatic gradients and patterns, further confusing interpretation.

In most modern classifications, four distinct deserts are recognized in the United States and northern Mexico (fig. 10.1).[3]

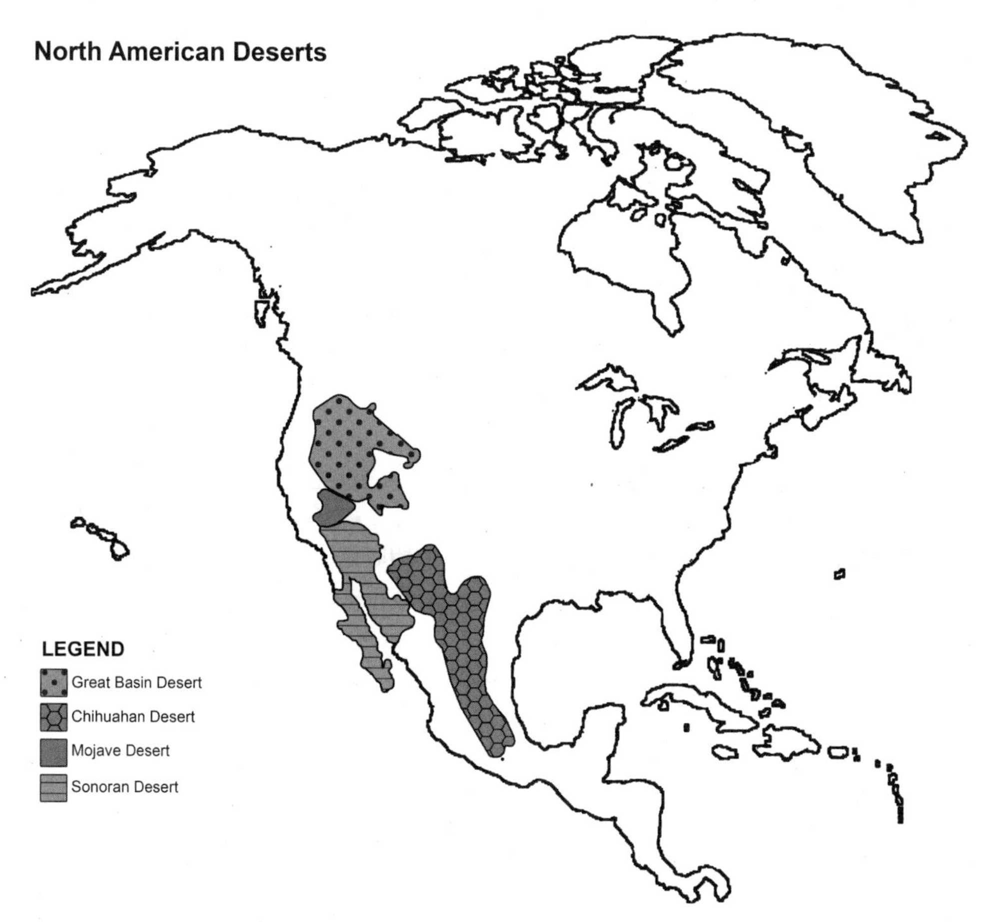


FIGURE 10.1 Deserts of North America.

**TABLE 10.1**

*Distinguishing characteristics of the four major deserts of North America*

| Characteristics | Great Basin | Mojave | Sonoran | Chihuahuan |
|---|---|---|---|---|
| Vegetation dominants | Giant sagebrush, salt grass, winter fat, snakebrush, saltbush | Creosote bush, Joshua tree, yucca, bur sage, and cacti *(Opuntia, Mammalaria, Echinocactus)*; grasses such as galleta, sand drop seed, and salt grass. | Mesquite, acacia trees, and shrubs, with creosote bush, agaves, yuccas, Joshua tree; many cacti, including saguaro. | Many grasses: drop seed, galleta, wheat grass; yuccas; some cacti (mostly *Opuntia*), bear grass, ocotillo (*Fouquieria splen dens*), creosote bush, and acacias. Few trees and shrubs. |
| Climate | Cooler year-round; subject to frost most of year; 4–11 inches of precipitation per year. Windy. Largely internal drainage and accumulation of salts in the soil. | Hotter than Great Basin, with similar but less predictable precipitation: 4–11 inches per year. Sparse vegetation. | Hot and dry: 0–13 inches of rain per year, mostly in winter in westernmost area. Occasional summer monsoon rains eastward, with reduced salinity because of increased rainfall. Includes rivers and dunes. | Hot, with cold winters (30°F average). Precipitation, mostly in summer: 3–20 inches. |
| Elevation (feet above sea level) | 4,000–8,000 | 500–4,000 | 0–3,500 | 3,500–5,000 |
| Location | Utah, Wyoming, southern Idaho, southeast Oregon, Nevada | California, southern Nevada | Southeast California, Baja, southwest and south central Arizona, northwestern state of Sonora, Mexico | New Mexico, southwestern Texas; Chihuahua, Nuevo Leon, and Coahuila, Mexico |

Distinctions are made primarily on the basis of vegetation composition (table 10.1). Plant communities in deserts, as in any ecosystem, are determined by the soils, seasonal distribution of available moisture, underlying geologic makeup, elevation, and disturbance history. Land use, particularly overgrazing by livestock, has been especially influential in bringing changes and obscuring the distinctions between North American deserts.

Three of the major deserts in North America, the Chihuahuan, the Sonoran, and the Mojave, are considered to be *hot deserts*, because of high summer temperatures. Most of the plant species of these deserts originated in the subtropics. The Great Basin Desert is considered a *cold desert*. Many of the plant species found in this desert have geographic distributions that overlap the other desert types. Nevertheless, each is distinct.

The classification of North American deserts is by no means universally agreed upon. For instance, some maintain that the Mojave is not a distinct desert but a transition between the Great Basin Desert and the Sonoran Desert. The Colorado Plateau is another large area of disagreement. This semiarid region of southern Utah and northern Arizona contains many protected natural areas, including Arches National Monument and the Grand Canyon. Some do not consider the Colorado Plateau a desert at all; others refer to it as the Painted Desert; still others consider this to be a southeastern extension of the Great Basin Desert. Even among those who agree upon four distinct deserts, there is disagreement over the geographic distribution. Some scientists argue that the distribution of animal species and climate should be considered along with distribution of plant species. Such details, however, are not critical to how you approach desert restoration.

The following are areas within the major desert types where you can visit to learn more. Most are labeled on highway maps (see the DesertUSA Web site: www.desertusa.com/). Be cautious about viewing these as reference examples. Most are highly disturbed, although they may contain some isolated areas that are in good condition.

- Arizona Upland Desert: The elevated portion of the Sonoran Desert in southern Arizona, characterized by saguaro cactus.
- Black Rock Desert: A subdivision of the Great Basin Desert located in northwestern Nevada just northeast of Pyramid Lake.
- Borrego Desert: The portion of the Sonoran Desert area just west of the Salton Trough of southeast California.
- Colorado Desert: The California portion of the Sonoran Desert west of the Colorado River.
- Escalante Desert: A subdivision of the Great Basin Desert just west of Cedar Breaks in southwestern Utah.
- Great Sandy Desert: A subdivision of the Great Basin Desert located in southeastern Oregon.
- Magdalena Desert: The Sonoran Desert on the lowest third of the Baja Peninsula.
- Painted Desert: This term is used differently by different writers. (1) A narrow desert strip running west of the Grand Canyon, north to south along U.S. Route 89, then turning east along Interstate 40 to just beyond Petrified Forest National Monument. (2) The entire region from the northern boundary of the Sonoran Desert of Arizona to southwestern Colorado and southern Utah, en-

compassing the Colorado River, the Colorado Plateau, and its numerous parks and monuments.

- Red Desert: The semiarid region of southwestern Wyoming, sometimes considered an extension of the Great Basin Desert.
- Smoke Creek Desert: A subdivision of the Great Basin Desert located in northwestern Nevada abutting the north end of Pyramid Lake.
- Trans-Pecos Desert: The Chihuahuan Desert west of the Pecos River in Texas.
- Yuma Desert: That portion of the Sonoran Desert just east of the Colorado River near Yuma, Arizona.

## Unique Biological Adaptations

Desert plants are adapted to the intense light and typical hot, dry summer conditions. *Succulence*, swollen stems, leaves, and roots, is an adaptation that increases the ability to store water and regulate temperature. During dry periods, some plants drop their leaves or become desiccated, often appearing lifeless. Some desert plants also have smaller leaves, or no leaves at all. Others, particularly annual plants, germinate, grow, and flower after brief rainy periods, surviving through dry seasons as seeds. Spines and needles, reflective epidermis, or profusely hairy leaves and stems reduce exposure to wind and sun and evaporative moisture loss. Some have thick, waxy cuticles to reduce water loss. Roots of many desert species form *distribution nets* to capture precipitation just below the surface. Creosote bush, among others, has remarkably deep roots to tap groundwater.

Cacti, along with many other desert succulents such as agaves, have a unique form of photosynthesis called *crassulacean acid metabolism*. These plants open stomates primarily at night when transpiration is relatively low and store carbon dioxide as a metabolic acid for photosynthetic use during the day. This enables them to carry out photosynthesis with stomates closed. Many of the grasses in hot deserts have a modified photosynthetic pathway that is more efficient at high temperatures.

Survival requires fine-tuned adaptations, so it is no surprise that desert animals are especially sensitive to vegetation and land-use changes. All have adapted for water conservation and regulation of their body temperatures. Burrowing underground to escape heat during the day is common. Indeed, many species forage primarily at night. Bearing of young often coincides with precipitation or cooler seasons, as with the spade-foot toad. Many animals, even mammals, primarily rely on metabolic water.

Although desert ecology is well documented, the history of changes in these arid and semiarid lands is not.[4] The degradation, however, has resulted from fairly consistent stressors.

## Stressors and Impacts

As with all ecosystems, desert restoration begins by a careful survey of conditions, with particular attention to stressors. The most common ones impacting desert ecosystems are discussed here.

**Overgrazing.** Most North American deserts have suffered from cattle and sheep grazing. Indeed, the expansion of deserts into adjacent semiarid ecosystems has resulted primarily from overgrazing. Valley bottoms and riparian areas in deserts are especially vulnerable. (See chapter 9 for further discussion of riparian restoration.) Grazing breaks up soil crusts, which primarily comprise cyanobacteria and lichens (moss and *Selaginella* in some locations), stabilizing soils against wind and water erosion and reducing evaporative water loss from underlying soils. An overabundance of nitrogen from feces and urine in areas where livestock congregate favors invasive plants. These changes favor larger woody plants with deeper roots, or short-lived annual weeds such as cheatgrass. Cattle, sheep, and goats preferentially eat bunch grasses and palatable native forbs, soon depleting some species and opening opportunities for invasive weeds and plants such as *Opuntia* cacti and mesquite.

**Fire.** We often do not think of fire in conjunction with deserts because overgrazing has resulted in reduction of fuels that historically carried periodic fire. Deterioration of soils and reduction of native grasses, concurrent with an increase in nonnative and woody vegetation, have also altered the fire regime. Mesquite and sagebrush, for example, have increased in the Great Basin Desert coinciding with a decrease in bunch grasses. Their spread, along with soil erosion and disturbance from grazing and even by off-road vehicles, has led to further reduction in the native grasses and forbs.[5] Prior to overgrazing, light fuels carried fire in many deserts. In some settings, fire may be a useful restoration tool to reduce woody vegetation and invasive species such as cheatgrass or Russian thistle.

**Erosion.** With overgrazing, wind and water erosion have increased, with resulting loss of finer particles from the surface of soil and an increase in desert pavements. In many deserts, erosion has been exacerbated by off-road vehicle use, and slow growth of vegetation makes recovery difficult. Deserts are especially vulnerable to rapid runoff. With greater runoff, drainageways are prone to erosion, with corresponding deepening of channels and bank erosion.

**Invasive plants and animals.** Cheatgrass gives many areas of the Mojave and Sonoran deserts an early spring "green flush," but they become nearly barren a month later. Cheatgrass is a widespread annual invasive grass throughout many degraded deserts. Weeds such as Russian thistle and other tumbleweeds have become nearly synonymous with deserts. Livestock avoid plants such as locoweeds and cacti. Locoweeds contain toxic levels of selenium.

**Water depletion.** In general, nonnative invasives use more water than the native species that are specially adapted to desert conditions. Increase of invasives together with soil erosion and compaction means that less water remains in the land. Nearly all disturbances in deserts lead to increased loss of water. The corollary is also worth noting: restoration can lead to increased water conservation. Even new springs sometimes emerge, and a return of perennial streams can result from control of invasive woody plants. Dense thickets of invasive species such as introduced willows, salt cedar, and Russian olive along waterways in deserts are also associated with the decline of native shoreland and wetland species.[6]

**Salinity.** Irrigation is often a threat to desert ecosystems. Because there is inadequate precipitation to flush salts from soils, irrigation results in accumulation of mineral salts as the water evaporates at the soil surface. Irrigation can also result in accumulation of toxic heavy metals—selenium, arsenic, cadmium, among others. These metals can affect nervous systems and survival of wildlife. Millions of desert acres worldwide are becoming poisoned by irrigation, creating especially challenging restoration.

There are naturally occurring specialized plant communities found in poorly drained saline flats and basins. For example, Nutall's saltbush and greasewood grow with as much as 80 to 90 percent salt. Even giant sagebrush survives in 30 percent salt. These natural communities need to be recognized and given special attention in desert restoration projects.

## Restoration Strategies

Some desert areas, even in sandy or rocky substrates and along drainageways, have retained native species in the seedbank. Heavily grazed or eroded areas, however, usually have few native species remaining. In preparing for desert restoration, first ascertain the status of a remnant seedbank. In your early assessment, determine where species need to be reintroduced, and where invasive plant challenges are likely.

Natural seeding of some desert plants can occur by birds, or by mammals such as pack rats that collect and hoard seeds. Protection from heat and desiccation, sufficient for establishment of species is critical. Seeding is primarily done during cooler and more available moisture conditions. Success can be increased by mulching to protect and anchor seeds and soil against desiccating winds. The following methods have proven useful.

**Restoration of drainage features and control of erosion.** In eroding drainageways, a useful strategy to keep moisture on the land involves the installation of stable rock grade-controls or brush wattles, which will capture sediments. The elevation of the bottom of a drainageway can be built up several feet, and in some projects generations of grade controls placed one on top of another slightly offset up or down the channel can result in restoring much of the original feature, especially in smaller drainageways. Grade controls, especially when combined with reintroduction (if necessary) of native vegetation, provide stabilization against further down-cutting. Native vegetation, such as deer grass, is easily established once stabilization begins and more reliable moisture is available in the rooting zone. As the channel fills, the seasonal water table is elevated, and this encourages the growth of plants on the adjacent land. Restoration of riparian and drainage features create refugia from which other tributary areas of the desert can be restored.

Restoration often requires physical intervention if mulch or simple grade controls are insufficient to stop the ongoing erosion. It may be necessary to bring in heavy equipment to reconstruct the landscape by backfilling gullies. In drainageways with larger tributary watersheds, reducing water flow by restoring uplands should precede treatment of drainageways.

Old two-track jeep roads and trails, as well as livestock trails, commonly become eroded gullies. These can be backfilled and regraded, followed by planting. Keeping livestock off such areas once they are planted, and creating water bars to divert or diffuse concentrated runoff will be essential. Fencing is almost always required to manage livestock.

Unless you are handy with heavy equipment, selecting the right operator to work with the restoration is essential. We have found that time is well spent in locating a conscientious operator, not just the guy down the road with a bulldozer. This is especially the case if you are not going to be present during the entire earthmoving operation. On some projects where we could not find an operator we trusted, we rode on the dozer during the entire job to ensure the work was properly done. One of the most important differences between a good operator and that guy down the road is his/her mindset. If you ask someone to fill an erosion gulley, a typical operator will simply look to directly adjacent lands for the fill. Fill should be taken from another disturbed area, and disturbance should not stray from the limits of the construction site you define. It can be remarkably difficult for some operators to understand that the shortest line between points A and B may not be the desired route. Their attitude can be similar to the off-road users who run their vehicles across sensitive lands.

During respread or grading, pay very close attention to soil compaction. If the soils are wet and have more than 10 percent clay content, be aware of the potential for increased compaction. We typically ask a dozer operator to compact the underlying materials, such as in a backfilled gulley, but then spread new topsoil over the area to a half-foot precision to avoid repeated passes over the new surface. We do the final seedbed preparation with small tractors or by hand to minimize compaction. Earthmoving is a last resort, as it is expensive and intrusive, and the costs for simple hand seeding and mulching can become overshadowed by tens of thousands of dollars per acre, if much earthmoving is required. Thus, earthmoving is generally done on a small, local scale.

The details of creating grade controls and water bars, or how to backfill a gulley, or reshape a drainage route, are beyond the scope of this book. Many guidebooks are available, including various USDA Forest Service and National Park Service road construction and erosion control manuals. Some specialized books on this subject are recommended.[7] This work is often not as simple as it seems. Become at least knowledgeable enough to oversee a contractor and work closely with the operator to communicate your goals clearly.

**Reintroducing desert crust.** Desert crusts, also called cryptobiotic crusts or biotic desert crusts, are created by lichens, mosses, fungi, cyanobacteria, and blue-green algae that bind soil particles together.[8] This crust forms an important, yet vulnerable protective layer over the soil. In disturbed areas, the soil often is crusted, but from compaction or sediments deposited by water. Whereas the natural desert crust absorbs water, the crust that forms as a result of disturbance forms a barrier to water, and increases runoff and erosion. The biotic crust seals in moisture, enabling better germination and growth of plants. Some of the more sensitive desert species cannot survive without the desert crust.

We have experimented with the restoration of other types of crusts with some success, and the same techniques may be useful and could be tested in restoring desert crusts. We learned that in some types of environments it is possible, especially at small scales, to restore bryophyte and lichen crusts. One method uses locally collected bulk samples of lichens and moss from the same type of habitat being restored (e.g., desert pavement or fine-soil habitats). While avoiding rocks, it is best to include some of the associated organic or mineral substrate. Place the collected plant tissue and associated substrate in a blender with buttermilk and puree the contents with a couple of quick bursts. The proper result will be finely dispersed fragments of plant tissue in the buttermilk. Collecting lichen specimens can be much more difficult than mosses. It may be necessary to scrape lichens from rocks if you cannot find enough on soil surfaces to generate sufficient material. Avoid disturbing pristine areas. We collected lichens in construction zones where the soil was scheduled to be moved.

We have sprayed the mix directly onto the surface of the exposed soil to establish new populations of moss and lichen crusts successfully. It is best to do this in the spring, during a cooler period. Typically, within a week or so, a mold-like growth develops on the soil. Within four to six weeks, small moss and lichen plantlets were present in our experiments.

Salvaging desert soils from where new roads are being constructed is a good source of soil for filling gullies, and the soil may also contain soil fungi needed for soil crust mother stock. Salvaged soils to be used for desert crusts or mycorrhizae should not be piled for long, as composting can kill the fungi. Also, it is not desirable to mix surface and subsoils. Pay attention to the seasonality of the salvage operation. Also be aware that dozer operators cannot strip topsoil with precision. They also may be challenged to spread a thin veneer evenly over a broad surface, so be prepared to do some handwork to redistribute salvaged soils.

**Seeding and seedbank stimulation.** Seeds must be protected from exposure to wind and sun, which can desiccate seedlings during germination, killing plants before they have a chance to become established. Also, wind or water erosion can blow or wash seed away. In deserts, only a very small percentage of seeds germinate in any year. Many become ant food or die, so protecting seeds through germination and establishment of seedlings is critical.

When seeding desert restorations, first consider the soil. Adjust the soil crust as described earlier, if necessary. If the surface or subsurface is badly compacted, as is often the case with overgrazing, it must be broken up prior to replanting. Breaking up subsurface compaction is most easily accomplished using a tractor and a heavy-duty chisel plow. Typically, chisel plow to depths of twelve to thirty-six inches, depending on the depth and severity of the compaction. For breaking up surface compaction, a useful tool is a roller with protruding spindles that punches indentations into the compacted layer every few inches. These indentations form small water-holding features and prevent seeds from being blown away. Broadcast the seed over the treated area. Seed will blow or wash into the indentations, which form microsites that protect seeds during germination. They also can collect enough moisture from even a single rainfall to support germination. This is a better technique for soil preparation than tillage, which exposes soil to erosion, damages the soil crust, and increases evaporative moisture loss.

On small areas, compacted soil can be broken by pulling a tined drag behind an ATV. Especially if netting or mulch can be secured over the area afterward, this method works well. Seed directly into the lightly worked soil surface before applying netting or mulch. An alternative mulching technique suitable for limited areas, such as old roadbeds, gullies, or trails, is to scatter brush over the area. This serves to capture blowing soil and seeds, but it also can prevent livestock from trampling the newly seeded areas. It also creates small microsites for the establishment of plants.

Compacted soil surfaces also can be broken by livestock if carefully managed. The animal's hooves will break up the hardened surface. Temporary fencing may be necessary to confine the livestock in desired areas for a few days. Seeds can be broadcast over the worked soil even while livestock are present. The stock will aid in working the seed into the soil but then should be removed before germination begins.

In suitable terrain, no-till drills can be used, even in desert pavement if pulled very slowly to avoid damage to the equipment. Drills will deliver seeds that need to be buried deeper, although the same equipment can be used to "dribble" seed on the surface simply by removing the feeder hoses and allowing them to dangle. Disks on the drill will create small furrows six to twelve inches apart, providing the microniches for the seeds that are dropped directly on the ground. They will then blow or wash into the grooves.

The timing of seeding is critical in a desert planting. Even though some deserts are hot during the day, all cool down at night. Many native plant seeds need to experience the cycling between cold and warm to germinate. Some need to nearly freeze for a few days or weeks before they will germinate. Most important, however, is precipitation. Although seeds of native species can remain viable for months or years, the longer they remain ungerminated, the more likely they will be moved by wind or destroyed by birds or ants.

To achieve optimal success, plan to seed in autumn so the seed has good soil contact and experiences the stratification provided by low temperature. Optimal germination conditions will then occur with winter or spring rains.

Western deserts have considerable elevation variability that affects the timing of germination and growing season. In exposed and elevated areas on the landscape, particularly on sandy or gravel surfaces, there is an effective reduction in available moisture because of the exposure. On steeper south- and west-facing slopes, moisture becomes limited earlier, often even before the growing season starts. In desert and similar environmental settings, such as reclaimed mined lands, seeding should be done in the winter, late fall, or very early spring. When possible, time the seeding to take advantage of precipitation during late winter and early spring.

If there is sufficient moisture, it may be possible to use cover crops that can germinate to stabilize soils in autumn and early spring. Cover crop species that can create a quick green-up after a late-autumn rain include winter rye, canola, and even annual rye grass. Seeding density should provide a scattering of cover crop plants and not a dense blanket that will rob moisture from the native plants you are trying to establish. Typically, cover crops such as winter rye can be seeded at 20–40 lb. per acre, canola at 1–2 lb. per acre, and annual rye grass at 15 lb. per acre to achieve sufficient cover to stabilize and lightly shade the surface. If cover crops can be established only in very early spring, consider using Italian millet drilled or broadcast at 10–15 lb. per acre with the native seeds, or overseeded after the native seeds have been sown.

Native seeds will be expensive, if they are even available. For a good description of how to harvest, process, store, and prepare your own locally collected native seeds, see Bainbridge's book.[9] Typically, we opt to seed native grasses at rates of 8–15 lb. of pure live seed (PLS) per acre when drilling, and slightly more when broadcasting. Native forb and shrub seeds are typically seeded at rates of 2–5 lb. of PLS per acre. To reduce cost, you can use the *strip introduction* strategy, especially if insufficient seed is available for uniform coverage. This involves strip plantings across landscape that, once established, will disperse into unplanted strips. Cover crops can be sown in the unseeded strips to build soil and break up compacted substrates. The added organic matter they provide may hasten the spread of native plants from the seeded strips. For example, strips of 25–100 feet in width can be spaced 100 feet or so apart. Seed additional strips in subsequent years between those first established.

We avoid planting strips in parallel bands because they can remain conspicuous for years. The aim is to create patterns of vegetation that conform to the landforms, diversity as quickly as possible. We jokingly suggest "getting the tractor driver liquored up so he or she can't drive a straight line" when laying out strips. The natural-looking pattern is most easily achieved by following the pattern of drainageways, or laying out strips that respond to the pattern of rock outcrops or other topographic features. During the restoration design process, discussed in chapter 3, find reference areas that will guide the patterns on the land. Often, there are features such as rock formations or topography that will produce natural patterns.

**Mulching.** Mulching is often essential in establishing desert plantings. Straw is an expensive but effective material that is easily applied with either straw crimpers (expensive implements you can purchase or lease) or farm disks set with the blades aligned (not offset, as when the disk is being used for soil tillage). Crimping secures the straw, which reduces evaporative moisture loss, adds organic matter to the soil as it decomposes, reduces erosion, and reduces heat absorption. Weed-free straw is more expensive but worth it considering the expense and time required to eradicate invasive species. Other mulches can be used. If available, brush can be piled on the surface. Brush can prevent or deter livestock disturbance, as well as protect the surface from excessive wind and drying. Netting, bark mulch, or wood chips are also effective but not economical except for applications in very small areas.

**Invasives.** Invasive plants are a pervasive challenge in most desert restorations. Many of the invasives are aggressive, persistent, and quickly surpass and suppress native species. Invasives are of two kinds: woody perennials, and herbaceous forbs and grasses that may be annuals, perennials, or biennials. The latter includes cheatgrass, red brome, cranes-bill, Russian thistle, and tansy mustard, among others. Drainageways, which may be intermittent or perennial, are where many plant and animal species live in deserts. Drainageways are also the locations where many invasive plants are likely to be established. Salt cedar, introduced willows, and Russian olive are especially prone to invading drainageways. These are aggressive sprouters, so cutting alone will not get the job done. Chinese elm is another species that moves across some uplands, often with aggressive native species such as mesquite, several *Acacia* species, several species of *Opuntia* cacti, large sagebrush, broom weed, and snakeweed. Cut invasive woody plants and paint stumps with 15–20 percent glyphosate. Glyphosate at lower concentrations (follow labels on containers) can be applied directly to herbaceous species. Unless invasives are too widespread and established, they often can be controlled as native species are being established. In drainageways, for example, woody invasives can be cut and used for mulch over seeded areas. Properly timed prescribed fire can reduce cheatgrass and many other invasives. A light ground fire can kill the new crop of seed before it is released from the parent plants and also reduce ungerminated seed on the soil surface. Fire requires sufficient fine grass or leaf litter. Where it is used, it creates a seedbed for seeding by one of the methods described earlier. Burning can also be used at times in control of dense salt cedar. If fuel is sufficient, fire can top-kill the adult plants and consume a majority of the leaf litter on the ground. Salt cedar will soon resprout and will require a follow-up herbicide treatment to kill it. Consult local experts on the proper herbicide to use as a foliar application on sprouts, or as a basal bark application. Back-to-back treatments of burning followed by herbicide may provide better control, after which you can begin seeding native plants into the ashes.

Be wary of some commonly recommended "range improvement techniques" such as roller-chopping, root-raking, chaining, or bulldozing, as these can cause serious disturbance and lead to worsened conditions. Such techniques are widely advocated to control invasive trees and shrubs, but the aim is to improve forage for livestock rather than restore desert ecosystems. Where burning is not an option, use herbicide treatments to reduce woody plants. Be sure to use recommended and proven herbicides and carefully follow directions on the label.

**Salinized and toxic soils.** One of the more difficult desert restoration challenges is salinized soils. Salinization results from a buildup of sodium, potassium, magnesium, and carbonate salts, usually following years of irrigation, or where erosion has down-cut and intercepted shallow flows or seeps of groundwater. Before attempting to restore salinized soils, determine if the condition was created by humans. If the salinity is a naturally occurring concentration, the best approach is to restore species that are adapted to saline conditions. If a result of irrigation, or contamination from well drilling, tilling in alum (aluminum potassium sulfate) and lime (calcium carbonate) can help replace the sodium or other toxic minerals with calcium or potassium at the soil particle binding sites. This exchange may require several years with natural precipitation, or copious irrigation can be used to speed up the replacement and help flush the salts from the soil. Bear in mind, however, that irrigation must be sufficient to flush the soil. Too little only adds additional minerals to the surface as the water evaporates. This process can be very costly.

Look for nearby areas in which to find seed and plant starts to establish vegetation on salinized fields. Be sure to check on ownership and obtain permission. It will be much quicker and easier to revegetate saline sites with tolerant species than to flush toxic compounds or salts out. It is worth getting soils analyzed to determine salt or mineral compositions, and work only with species known to tolerate those conditions. There are several excellent regional publications that address salinized and toxic soil management, such as the *Western Fertilizer Handbook.*[10] There also may be local projects you can visit to learn techniques that have been successful, although most work has focused on maintaining agricultural productivity rather than ecological restoration.

**Wildlife, including invertebrates.** In deserts, many nocturnal mammals are agents for dispersal of plant seed and fruit. These animals are often reduced or eliminated by human disturbances such as agriculture and grazing. As desert restoration begins, do not be dismayed to see mounds of soils left in the wake of various rodents and insects, especially ants. These creatures gather and scatter seeds and should be encouraged by the restoration process. Their activities foster the diversity of plants.

## Conclusions

A great way to get ideas for restoration is by visiting protected desert natural areas. Many have interpretive museums, trails, and representative restoration projects. Do not overlook areas where wildfires have burned through desert. Response is often remarkable, with a reduction in invasive woody vegetation and cacti and the stimulation of native grasses and wildflowers. Even some ranchers are now using fire as a management tool with good results. There often is a night-and-day difference across fences between neighboring ranches, one of which uses fire and one that does not. The Buena Vista National Wildlife Refuge in Arizona, south of Tucson, is a great example of Sonoran desert restoration. During our first visit in the late 1980s, before the refuge began restoration, the land was dominated by farming and ranching. In only a decade, the transformation following restoration was striking, as bunch grasses, native trees, and shrub-lined meandering drainageways and seasonal wetlands were restored. Especially nearer the drainageways, dense bunches of deer grass and spring ephemeral and perennial wildflowers were scattered among native shrubs. Invasive trees such as Russian olive and salt cedar grew along drainageways, and mesquite in some old fence lines remained.

Some desert wilderness areas retain vestiges of the historic conditions and may provide the best models to emulate in developing restoration goals. The Pinacate Wilderness Area, Sonoran Desert World Heritage Bioreserve, in northern Mexico just south of Organ Pipe National Monument, Arizona, demonstrates both badly disturbed landscapes and remnants of high-quality desert. Many other high-quality natural areas, as well as degraded deserts, can be found, most after long drives over dusty roads. You will have no problem, however, finding numerous examples of depleted and scarred deserts. While deserts are foreign to most of us who live in the temperate world, they are fascinating ecosystems well deserving special attention.

# SPECIES LIST

| *Species common names* | *Scientific names and other taxonomy* |
|---|---|
| Ailanthus, tree-of-heaven | *Ailanthus altissima* |
| American basswood | *Tilia americana* |
| American beech | *Fagus grandifolia* |
| American bison, bison, American buffalo | *Bison bison* |
| American black currant | *Ribes americanum* |
| American bladdernut | *Staphylea trifolia* |
| American chestnut | *Castanea dentata* |
| American elm | *Ulmus americana* |
| American hazel, hazelnut | *Corylus americana* |
| American holly | *Ilex opaca* |
| American hornbeam, muscle-wood, hornbeam, ironwood | *Carpinus caroliniana* |
| American toad | *Bufo americanus* |
| American wild plum | *Prunus americana* |
| Annual ryegrass | *Lolium multiflorum* |
| Aspen (see quaking aspen, big-toothed aspen) | |
| Aster | *Aster* spp. |
| Bald cypress | *Taxodium distichum* |
| Basswood, (see American basswood) | |
| Beaked hazel | *Corylus cornuta* |
| Bergamot | *Monarda didyma* |
| Big bluestem | *Andropogon gerardi* |
| Big-toothed aspen | *Populus grandidentata* |
| Birch | *Betula* spp. |
| Birdsfoot violet, birds-foot violet | *Viola pedata* |
| Bison (see American bison) | |
| Bittersweet | *Celastrus scandens* |

Black cherry *Prunus serotina*
Black cottonwood *Populus trichocarpa*
Black locust, yellow locust *Robinia pseudoacacia*
Black mangrove *Avicennia germinans*
Black maple *Acer nigrum*
Black oak *Quercus velutina*
Black raspberry *Rubus occidentalis*
Black swallowtail *Papilio polyxenes asterius*
Black walnut *Juglans nigra*
Black willow *Salix nigra*
Black-footed ferret *Mustela nigrepes*
Blackjack oak *Quercus marilandica*
Black-nosed dace, blacknose dace, black nosed dace *Rhinichthys atratulu*
Black-tailed prairie dog *Cynomys ludovicianus*
Blazing star *Liatris* spp.
Blue gramma, blue gramma grass *Bouteloua gracilis*
Blue hyssop *Hyssopus officinalis*
Blue jay *Cyanocitta cristata*
Blue vervain *Verbena hastata*
Bluebell *Mertensia virginicus*
Blue-wing teal, blue wing teal, blue-winged teal *Anas discors*
Bobolink *Dolichonyx oryzivorus*
Bobwhite quail, northern bobwhite, Virginia quail *Colinus virginianus*
Boxelder *Acer negundo*
Broad-leaved cattail *Typha latifolia*
Brook stickleback *Culaea inconstans*
Brook trout, speckled trout *Salvelinus fontinalis*
Broom weed *Xanthocephalum dracunculoides*
Broomsedge *Andropogon virginica*
Buckwheat *Fagopyrum esculetum*
Buffalo grass *Buchloe dactyloides*
Bur oak *Quercus macrocarpa*
Burdock *Arctium* spp.
Burrowing owl *Athene cunicularia*
Butterfly weed *Asclepias tuberosa*
Butternut *Juglans cinerea*
Caddisfly, shadfly >1,200 North American species in the Tricoptera order
Canada thistle *Cirsium arvense*
Cardinal flower *Lobelia cardinalis*
Casuarina *Casuarina* spp.
Catbriar, cat-briar, cat briar *Smilax* spp.
Cattail (see broad-leafed cattail, narrow-leafed cattail)
Cheatgrass, cheat grass, downy brome *Bromus tectorum*

| | |
|---|---|
| Chestnut blight | *Cryphonectria parasitica* |
| Chestnut oak | *Quercus prinus* |
| Chinese elm | *Ulmus parvifolia* |
| Chokecherry | *Prunus virginiana* |
| Common elderberry | *Sambucus canadensis* |
| Common hemp nettle | *Galeopsis tetrahit* |
| Common reed grass | *Phragmites australis* |
| Common snipe | *Gallinago gallinago* |
| Common sucker, white sucker, brook sucker | *Catastomus commersonii* |
| Common wolfstail | *Lycurus phleoides* |
| Compass plant | *Silphium laciniatum* |
| Cordgrass, cord grass | *Spartina pectinata* |
| Cottontail rabbit, eastern cottontail | *Sylvilagus floridanus* |
| Cranesbill | *Erodium cicutarium* |
| Creeping muhly | *Muhlenbergia repens* |
| Crested wheatgrass, crested wheat grass | *Agropyron cristatum* |
| Crown vetch | *Coronilla varia* |
| Cup plant, cupplant | *Silphium perfoliatum* |
| Deer grass | *Sporobolus wrightii* |
| Dickcissel | *Spiza americana* |
| Dogwood | *Cornus* spp. |
| Douglas-fir | *Pseudotsuga menziesii* var. *menziesii* |
| Dowitcher | *Limnodromus* spp. |
| Dragonfly | Insects in the Odonata order, Epiprocta suborder |
| Eastern hemlock | *Tsuga canadensis* |
| Eastern meadowlark | *Sturnella magna* |
| Eastern red-cedar, eastern red cedar, juniper | *Juniperus virginiana* |
| Eastern white pine | *Pinus strobus* |
| Elk, wapiti | *Alces alces* |
| Engelmann spruce | *Picea engelmannii* |
| European buckthorn, common buckthorn (see glossy buckthorn) | *Rhamnus cathartica* |
| European carp | *Cyprinus carpio* |
| European earwig | *Forficula auricularia* |
| European fragile willow (see fragile willow) | |
| Evening primrose | *Oenothera biennis* |
| Fan-tailed darter, fantailed darter | *Catanotus flabellaris flabellaris* |
| Figwort | *Schropularia marilandica* |
| Fir | *Abies* spp. |
| Fox grape | *Vitis labrusca* |
| Fox sedge | *Carex vulpinoidea* |
| Foxtail grass | *Setaria* spp. |

| | |
|---|---|
| Fragile willow, European fragile willow | *Salix fragilis* |
| Garlic mustard | *Alliaria petiolata* |
| Giant ragweed | *Ambrosia trifida* |
| Giant sagebrush | *Artemisia tridentata* |
| Gipsy moth | *Lymantria dispar* |
| Glossy buckthorn | *Rhamnus frangula* |
| Golden alexander | *Zizia aurea* |
| Golden glow | *Rudbeckia lacinata* |
| Golden-glow sunflower | *Rudbeckia lacinata* |
| Goldenrod | *Solidago* spp. |
| Gopher | *Geomys* spp. and *Thomomys* spp. |
| Grape | *Vitis* spp. |
| Gray dogwood | *Cornus racemosa* |
| Gray tree frog | *Hyla versicolor* |
| Greasewood | *Sarcobatus vermiculatus* |
| Great blue heron | *Ardea herodias* |
| Greater prairie chicken | *Tympanuchus cupido* |
| Green ash | *Fraxinus pennsylvanica* |
| Green tree frog | *Hyla cinerea* |
| Ground squirrel | *Spermophilus* spp. |
| Hackberry, sugarberry | *Celtis laevigata* |
| Hairy-node sedge | *Carex trichocarpa* |
| Hawthorn | *Crataegus* spp. |
| Henslow's sparrow | *Ammodramus henslowii* |
| Hill's oak, northern pin oak | *Quercus ellipsodalis* |
| Honeysuckle | *Lonicera* spp. |
| Honeysuckle vine (see Japanese honeysuckle) | |
| Hophornbeam | *Carpinus caroliniana* |
| Horned lark | *Eremophila aplestris* |
| House wren | *Triglodytes aedon* |
| Indian grass, indiangrass | *Sorghastrum nutans* |
| Indigo bush, false indigo, bastard indigo, river locust | *Amorpha fruticosa* |
| Ironweed | *Veronia altissima* |
| Italian millet | *Setaria sibirica* |
| Ivory-billed woodpecker | *Campephilus principalis* |
| Jack pine | *Pinus banksiana* |
| Jackrabbit | *Lepus* spp. |
| James' galletta | *Pleuraphis jamesii* |
| Japanese brome | *Bromus japonicus* |
| Japanese honeysuckle | *Lonicera japonica* |
| Japanese knotweed | *Fallopia japonica*, syn. *Polygonum cuspidatum* |
| Joe-pye weed | *Eupatorium purpureum* |
| Kentucky bluegrass | *Poa pretensis* |

| | |
|---|---|
| Kudzu, kudzu vine | *Pueraria lobata* |
| Ladino clover | *Trifolium repens* |
| Laurel oak | *Quercus laurifolia* |
| Lead plant | *Morpha canescens* |
| Leafy spurge | *Euphorbia esula* |
| Leafy spurge flea beetle | *Aphthona* spp. |
| Leopard frog | *Rana pipiens* |
| Little bluestem | *Schizachyrium scoparium* |
| Loblolly pine | *Pinus taeda* |
| Locoweed | Primarily plants in the *Astragalus* and *Oxytropis* genera |
| Lodgepole pine | *Pinus contorta* |
| Longleaf pine | *Pinus palustris* |
| Marsh hawk, northern harrier | *Circus cyaneus* |
| Mayfly | About 630 North American species in the Ephemeroptera order |
| Meadowlark (see eastern meadowlark and western meadowlark) | |
| Meadowsweet, white meadowsweet | *Spiraea alba* |
| Mediterranean geranium | *Erodium cicutarium* |
| Melaleuca | *Melaleuca quinquenervia* |
| Mesquite | *Prosopis* spp. |
| Mockernut hickory | *Carya tomentosa* |
| Monarch butterfly | *Danaus plexippus* |
| Moonseed | *Menispermum canadense* |
| Mountain mint, common mountain mint | *Pycnanthemum virginianum* |
| Multiflora rose | *Rosa multiflora* |
| Nannyberry | *Viburnum lentago* |
| Narrow-leaved cattail | *Typha angustifolia* |
| Needle-and-thread grass | *Stipa comata* |
| Nettle (see stinging nettle) | |
| New Jersey muhly | *Muhlenbergia torreyana* |
| New Mexico feathergrass | *Stipa neomexicana* |
| Northern dropseed, prairie dropseed | *Sporobolus heterolepis* |
| Northern pike | *Esox lucius* |
| Northern red oak | *Quercus rubra* |
| Northern spotted owl | *Strix occidentalis caurina* |
| Norway maple | *Acer platanoides* |
| Nutall's saltbush | *Atriplex nutalli* |
| Oak wilt | *Ceratocystis fagacearum* |
| Opossum, Virginia opossum | *Didelphis virginiana* |
| Oregon ash | *Fraxinus latifolia* |
| Overcup oak | *Quercus lyrata* |
| Pack rat, trade rat, wood rat | *Neotoma* spp. |

| | |
|---|---|
| Pale purple coneflower | *Echinacea pallida* |
| Paper birch, white birch, canoe birch | *Betula papyrifera* |
| Pasqueflower, pasque flower | *Anemone patens* |
| Pennsylvania sedge | *Carex pensylvanica* |
| Persimmon | *Diospyros virginiana* |
| Phlox | *Phlox* spp. |
| Pignut hickory | *Carya glabra* |
| Pinyon pine | *Pinus edulis* |
| Pitch pine | *Pinus rigida* |
| Poison sumac | *Rhus vernix* |
| Ponderosa pine | *Pinus ponderosa* |
| Porcupine grass | *Stipa spartea* |
| Post oak | *Quercus stellata* |
| Prairie cordgrass | *Spartina pectinata* |
| Prairie crab, prairie crab apple | *Malus ioensis* |
| Prairie dog | *Cynomys* spp. |
| Prairie fame-flower | *Talinum parviflorum* |
| Prairie lily | *Lilium philadelphicum* |
| Prickly-ash | *Xanthoxylum americanum* |
| Privet | *Ligustrum* spp. |
| Pronghorn | *Antilocapra americana* |
| Purple loosestrife | *Lythrum salicaria* |
| Quack grass | *Agropyron repens* |
| Quaking aspen | *Populus tremuloides* |
| Question mark butterfly | *Polygonia interrogationis* |
| Raccoon | *Procyon lotor* |
| Rainbow darter | *Etheostoma caeruleum* |
| Raspberry (see red raspberry) | *Rubus* spp. |
| Red admiral | *Vanessa atalanta rubria* |
| Red alder | *Alnus rubra* |
| Red brome | *Bromus rubens* |
| Red clover | *Trifolium pretense* |
| Red elm | *Ulmus rubra* |
| Red fox | *Vulpes vulpes* |
| Red mangrove | *Rhizophora mangle* |
| Red maple | *Acer rubrum* |
| Red milkweed (see swamp milkweed) | |
| Red oak (see northern red oak) | |
| Red raspberry | *Rubus ideaus* |
| Redbud | *Cercis canadensis* |
| Red-cockaded woodpecker | *Picoides borealis* |
| Red-node sedge | *Carex trichocarpa* |
| Reed canary grass | *Phalaris arundinacea* |
| Ring-necked pheasant | *Phasianus colchicus* |

| | |
|---|---|
| River birch | *Betula nigra* |
| Rosinweed | *Silphium integrifolium* |
| Russian thistle, tumbleweed, wind witch | *Salsola kali* |
| Russian olive, Russian-olive | *Elaeagnus angustifolia* |
| Saguaro cactus | *Carnegiea gigantea* |
| Salt cedar, saltcedar, salt-cedar, tamarisk | *Tamarix ramosissima* |
| Salt-marsh grass | *Spartina alterniflora* |
| Sand dropseed | *Sporobolus cryptandrus* |
| Sandbar willow | *Salix interior* |
| Sassafras | *Sassafras albidum* |
| Scarlet oak | *Quercus coccinia* |
| Scotch pine, Scots pine | *Pinus sylvestris* |
| Sedge | Primarily members of the *Carex* genus |
| Shagbark hickory | *Carya ovata* |
| Sharp-tailed grouse | *Tympanuchus phasianellus* |
| Shooting star | *Dodecatheon meadia* |
| Short-eared owl | *Asio flammeus* |
| Shortleaf pine | *Pinus echinata* |
| Side-oats gramma, sideoats gramma | *Bouteloua curtipendula* |
| Silky dogwood | *Cornus amomum* |
| *Silphium* | A genus with about 20 species, in the Asteraceae family |
| Silver maple | *Acer saccharinum* |
| Sitka spruce | *Picea sitchensis* |
| Skipper | Butterflies in the Hesperiidae family |
| Skunk cabbage | *Symplocarpus foetidus* |
| Slash pine | *Pinus elliottii* |
| Smartweed | *Polygonum* spp. |
| Smooth brome | *Bromus inermis* |
| Smooth sumac | *Rhus glabra* |
| Snake weed | *Persicaria bistorta* |
| Sourwood | *Oxydendrum aboreum* |
| Southern red oak | *Quercus falcata* |
| Southern red-bellied dace, southern redbellied dace | *Phoxinus erythrogaster* |
| Spade-foot toad | *Scaphiopus* and *Spea* spp. |
| Speckled alder, gray alder, grey alder | *Alnus incana* |
| Spotted sandpiper | *Actitis macularius* |
| Spring peeper | *Pseudoacris crucifer* |
| Spruce budworm | *Choristoneura* spp. |
| Spruce | *Picea* spp. |
| Staghorn sumac | *Rhus typhina* |
| Steeplebush, hardtack, tomentose meadowsweet | *Spiraea tomentosa* |

| | |
|---|---|
| Stinging nettle | *Urtica dioica* |
| Subalpine fir | *Abies lasiocarpa* |
| Sugar maple | *Acer saccharum* |
| Sumac (see staghorn sumac, winged sumac, smooth sumac) | |
| Swamp chestnut oak | *Quercus michauxii* |
| Swamp milkweed | *Asclepias incarnata* |
| Swamp white oak | *Quercus bicolor* |
| Sweet black-eyed Susan, sweet black eyed Susan | *Rudbeckia subtomentosa* |
| Sweetgum | *Liquidambar styriciflua* |
| Switchgrass | *Panicum virgatum* |
| Sycamore | *Platanus occidentalis* |
| Tamarisk (see salt cedar) | |
| Tansy mustard | *Descurainia pinnata* |
| Tartarian honeysuckle | *Lonicera tartarica* |
| Thistle | Primarily members of the *Cirsium* genus |
| Three awn, threeawn | *Aristida wrightii* |
| Tiger swallowtail | *Papilio glaucus* |
| Timber wolf, gray wolf | *Canis lupus* |
| Tree swallow | *Tachycineta bicolor* |
| Trillium | *Trillium* spp. |
| Tulip-poplar, yellow-poplar, tulip tree | *Liriodendron tulipifera* |
| Tumbleweed (see Russian thistle) | |
| Turk's-cap lily | *Lilium superbum* |
| Turtlehead | *Chelone* spp. |
| Upland sandpiper | *Bartramia longicauda* |
| Valley oak | *Quercus lobata* |
| Violet | *Viola* spp. |
| Virginia bluebell | *Mertensia virginica* |
| Virginia creeper | *Parthenocisis quinquefolia* |
| Virginia pine | *Pinus virginiana* |
| Vole | *Microtus* spp. |
| Water hickory | *Carya aquatica* |
| Water oak | *Quercus nigra* |
| Water tupelo | *Nyssa aquatica* |
| Western chorus frog | *Pseudoacris triseriata* |
| Western hemlock | *Tsuga heterophylla* |
| Western meadowlark | *Sturnella neglecta* |
| Western prairie fringed orchid | *Platanthera praeclara* |
| Western red cedar | *Thuja plicata* |
| Western wheatgrass | *Agropyron smithii* |
| White alder | *Alnus rhombifolia* |
| White ash | *Fraxinus americana* |

| | |
|---|---|
| White mangrove | *Laguncularia racemosa* |
| White mulberry | *Morus alba* |
| White oak | *Quercus alba* |
| White-tailed deer | *Odocoileus virginianus* |
| Wild parsnip | *Pastinaca sativa* |
| Wild rose | *Rosa blanda* |
| Willow | Trees and shrubs in the *Salix* genus |
| Winged sumac | *Rhus copillinum* |
| Wiregrass, pineland threeawn grass | *Aristida trisetum* |
| Wood duck | *Aix sponsa* |
| Woodchuck, groundhog | *Marmota monax* |
| Yellow coneflower | *Ratibida pinnata* |
| Yellow hyssop | *Agastache nepetoides* |
| Yellow lousewort, wood betony | *Pedicularis canadensis* |

## GLOSSARY

**allelopathy.** The inhibition of growth of a plant due to metabolites released by another plant of the same or different species. When these metabolites are released into the environment, they can inhibit the development of neighboring plants.

**alluvial soil.** A fine-grained fertile soil deposited by water flowing over floodplains or in river beds.

**anoxic.** Greatly deficient in oxygen, usually in reference to zones in water bodies.

**anthropogenic.** Having an origin that was influenced or shaped by humans.

**anthropogenic feature.** A feature of the landscape derived from human activities.

**basal area.** The sum of the cross-sectional areas of all tree stems in a unit area, usually expressed as square feet per acre, or square meters per hectare. This is the most commonly used metric for estimating biomass of trees.

**biennial species.** A flowering plant taking two years to complete its life cycle. In the first year, the plant grows leaves, stems, and roots. During the second year, the plant flowers, produces fruit and seeds, and then dies.

**bioaccumulation.** An increase in the concentration of a chemical in specific organs or tissues at higher than expected concentrations, or at successively higher concentrations in higher tropic levels of a food chain.

**biofiltration.** A pollution control technique using living vegetation and microorganisms to capture and hold or biologically degrade pollutants.

**biological integrity index.** A summary numeric that can be used to compare ecosystems, based on the number of different types of groups and species present. Areas such as streams with more types of groups of macroinvertebrates would, for example, have a higher index score then the macroinvertebrates found in an agricultural ditch.

**biomass.** The total mass of living matter within a square meter or hectare of any given habitat, or the mass of an organism.

**biome.** Terrestrial regions inhabited by certain types of life, especially vegetation. Examples include boreal forests and short grass prairies.

**bioremediation.** Any process where microorganisms, fungi, plants, or their enzymes are used to metabolize contaminants and restore polluted soil or water.

**brush wattle.** A rough bundle of brushwood used to strengthen an earthen structure, typically used in stream bank stability and restoration.

**caliche.** A white-to-gray irregular accumulation of calcium carbonate in soils of arid regions.

**carbon credit.** Carbon credits are created when one party is able to reduce its greenhouse gas emissions below some required regulated level. The quantity below a required level can then be exchanged (sold) to entities needing credits.

**carbon sink.** A naturally occurring system, such as soil or water, in which carbon dioxide can be assimilated or accumulated and sequestered in long-term storage, such as organic matter in soils, carbonates in ocean water, or wood in forest trees.

**carrying capacity.** The maximum number of individuals of a species an ecological unit can sustain as constrained by available food, water, and energy. For example, the number of cows that can be sustained on an acre of pasture, or deer in a unit of habitat.

**connectivity.** The way in which landscape or habitat features are connected by similar or dissimilar habitat.

**conservative species.** A species with a very narrow tolerance for certain conditions, typically not able to tolerate human-caused disturbances. They often, therefore, are found in less common habitats.

**contour farming.** A farming practice of tillage across a slope, following the elevation contour lines, thus reducing soil erosion and increasing water retention.

**cover crop.** A temporary vegetative cover that provides protection for the soil, preventing erosion and adding nutrients to promote the establishment of a permanent vegetative layer.

**cut-banks.** An erosion feature of a stream. Cut-banks are found along the outside of a stream bend.

**cyanobacteria.** Single-celled, prokaryotic, microscopic organisms. Before being reclassified as *monera,* they were called blue-green algae.

**discharge rates.** The volume of water passing through a channel during a given time, typically measured in cubic feet per second (cfs).

**disturbance.** A natural or anthropogenic event that changes the structure, content, and/or function of an ecosystem, usually in a substantial manner (also called **perturbation**). Alternatively, one incremental event in a sequence that causes ecosystem degradation. Many ecosystems sustain and require regular disturbances to maintain diversity, structure, functions, and use by plant and animal species (e.g., fire in jack pine forests; relationships between moose and suckering aspen; prairie ecosystems; grassland ecosystems).

**dormant planting.** Drilling or broadcast seeding after cold weather arrives so seed does not germinate until the following spring.

**dowsing.** See **witching.**

**drainageway.** Nonnavigable, aboveground watercourses.

**ecological resiliency.** The capacity of ecosystems to recover or shift to a qualitatively different state that is controlled by a different set of processes following disturbance.

**ecological restoration.** The process of assisting the recovery of an impaired ecosystem.

**ecosystem functionality.** The processes resulting from interacting elements of an ecosystem: habitat and community structure, biodiversity, and biogeochemical cycles.

**ecosystem health.** State or condition of an ecosystem in which its dynamic attributes are expressed within "normal" ranges of activity, relative to its ecological stage of development.

**ecosystem-scale strategy.** Addressing the physical, biological, or chemical-structuring drivers or constraints that control the ecosystem at the scale at which they operate.

**ecotone.** A transitional zone between two ecological communities or ecosystems, typically with some species common to both.

**ephemeral.** Temporary, or very short lived, such as spring wildflowers of the deciduous forest, or seasonal ponds.

**eutrophication.** Physical, chemical, and biological changes that take place after a lake, an estuary, or a slow-flowing stream receives inputs of plant nutrients—mostly nitrates and phosphates—from natural erosion and runoff from the surrounding land basin.

**extirpated species.** A species that ceases to exist in one area, but still exists elsewhere.

**fen.** A wetland created by seepage, generally wet the entire year.

**frost-heaving.** The process whereby a mass of sediment or soil undergoes freezing, expansion, and actual lifting, followed by thawing, contraction, and lowering of the mass.

**gap-phase.** Tree species adapted to quickly grow into small openings that develop in the forest canopy.

**genetic heterogeneity.** Variation in gene frequency among members of a gene pool or species.

**georeferencing.** The use of a geographic positioning system (GPS) to map location.
**glacial till.** Drift (earth and rocks) that is deposited directly from glacial ice and therefore not sorted. Also known as **till** or **glacial drift.**
**green manure.** Freshly cut or still-growing green vegetation that is plowed into the soil to increase the organic matter and humus available to support crop growth.
**ground-truthing.** Confirming, on the ground, the interpretation of conditions and deductions made through the use of remote sensing, such as aerial photography.
**growth ring.** A layer of wood formed in a plant during a single period of growth.
**grub.** A stunted tree caused by repeated top-kill by fire, winter desiccation, or browsing. Some insect larvae are also called grubs.
**guild.** A group of organisms that are part of the same "functional group," especially those sharing similar habitat, food, or foraging habits.
**impervious surface.** Materials that greatly retard or prevent movement of fluid through them.
**indicator species.** A species whose presence and/or abundance indicates conditions in a specific habitat, community, or ecosystem.
**infiltration.** The process by which water on the ground surface enters the soil.
**integrity.** State or condition of an ecosystem that expresses attributes of biodiversity (e.g., composition and structure) within "normal" ranges, relative to its ecological stage of development.
**keystone species.** A species of plant or animal that produces a major impact (as by predation or habitat modification) on its ecosystem and is essential to maintain optimum ecosystem function or structure.
**ladder fuel.** Flammable material that can carry ground fire into the canopy of a forest.
**levee.** A natural or artificial slope or wall that prevents flooding of the land behind it.
**loess.** Windblown silt and clay deposits derived from deserts, glacial outwash, or floodplains.
**macroinvertebrate.** Invertebrates visible to the unaided eye usually associated with aquatic habitats.
**marl.** Unconsolidated clays, silts, sands, or mixtures of these materials that contain a variable content of calcareous material.
**meander belt.** The zone along the floor of a valley through which a meandering stream periodically shifts its channel.
**microtopography.** A feature on the landscape created by individuals or groups of plants and animals. Examples include crayfish burrows, badger burrows, and wind-thrown tree mounds.

**mottled.** Irregular spots and streaks of different colors found in imperfectly drained soil where a portion of the iron is reduced (gray).

**muck.** An organic soil composed of highly decomposed organic material.

**mycorrhiza association.** A mutualistic relationship between a fungus and the roots of a plant in which both benefit. The fungus greatly extends the ability of the plant to extract water and nutrients from the soil.

**naturalization.** A process of near abandonment or intentional acceptance of default plant and animal communities that develop in human-altered systems, typically resulting in human-made ecosystems with a combination of native and nonnative species that become established, managed, or retained.

**net primary productivity (NPP).** The accumulation of energy or biomass in excess of the energy used for respiration. NPP is closely approximated by the increase in total biomass.

**niche.** The total way of life or role of a species in an ecosystem. It includes all physical, chemical, and biological conditions a species needs to live and reproduce in an ecosystem.

**nitrogen fixation.** The assimilation of free nitrogen from the air by microorganisms, indirectly making the nitrogen available to plants.

**oldfield.** An abandoned agricultural field or pasture, usually one that was once tilled.

**outwash.** Gravel, sand, silt, and clay deposited by the meltwater discharging from a glacier's terminus.

**perennial stream.** A stream or river that has continuous flow in parts of its bed year-round during years of normal rainfall.

**perturbation.** A disturbance that results in departure of a natural system from a normal state or trajectory.

**pH.** A numeric value that indicates the concentration of H ions in soil or water; can be thought of as relative acidity or alkalinity. The pH scale varies from 0 to 14, with the neutral point (neither acid nor alkaline) at 7. Acid solutions have a pH value lower than 7, and basic or alkaline solutions have a pH value greater than 7.

**phenology.** The study of the times of recurring natural phenomena.

**pocosin.** A perched wetland forming as a result of water seeping from the ground, with buildup of acidic peat and associated vegetation. Pocosins are most common along the southeastern U.S. coastal plain.

**primary production.** The accumulation of energy by a plant or group of plants in a defined area, typically measured by accumulation of biomass. (See **net primary production.**)

**propagule.** Any plant structure capable of plant propagation.

**recharge zone.** The region where water enters the ground to contribute to the zone of saturation or groundwater.

**reference area.** The reference may consist of one or more intact and functioning ecosystems that the impaired ecosystem is expected to emulate when it is restored or rehabilitated. Reference site or descriptions of such ecosystems that are desired in the restored or rehabilitated ecosystem (reference model).

**reference reaches.** A river segment that represents a stable channel within a particular valley morphology, used as a point of reference for the restoration of a similar segment of stream.

**resiliency.** The ability of a living system to recover conditions after being exposed to a disturbance or perturbation.

**rill erosion.** Erosion by running water that scours small channels into the ground.

**rip-rap.** Layers or assemblages of broken stones or unnatural materials placed to protect an embankment against erosion by running water or breaking waves.

**scarify.** Cutting, scratching, or softening of the seed coat to hasten germination.

**seedbank.** The reserves of viable seeds present in the soil and on the surface.

**sheet erosion.** Erosion that is more or less evenly distributed over the surface and removes thin layers of soil.

**shrub carr.** A wetland community dominated by tall shrubs.

**stage.** Water depths, such as in a flood, but may reference stream flow during subflood conditions.

**steppe.** Short-grass prairie or semidesert with a mixture of herbaceous and shrubby species.

**stratification.** Pretreating seeds to simulate winter conditions so germination can occur.

**stream profile.** Change in elevation over a distance in a stream channel.

**stream terrace.** Relict floodplain, from periods when a stream was flowing at a higher elevation.

**subsoil.** Soil that occurs beneath the topsoil or A horizon.

**surface drainage divides.** The ridges or highest points between neighboring watersheds or drainage basins.

**suspended load.** Sediment in a stream or river carried by the water.

**sustainability.** The continuation of a system such that resources, including its diversity, are not depleted or damaged. The capacity of a system to function indefinitely through various conditions where stocks of renewable resources are not used faster than they are renewed, and no wastes are generated.

**sustainable development.** Development that meets the needs of the present and future without compromising either.

**terracing.** The creation of less-steep surfaces by construction of embankments parallel to the slope or contour of the land, at specific intervals.

**terrestrial ecosystem.** Organisms and their environment that occur on land, as opposed to those in water.

**turbidity.** Refers to the clarity of water, which is a function of how much material is suspended in the water.

**water bar.** A constructed feature in streams used to control down-cutting of the channel; typically created from fallen trees.

**watershed.** A land area that contributes water to a particular stream or river system.

**wetland delineation.** Establishing a boundary between the upland community and wetland community.

**wickiup.** A domed single-room dwelling used by certain Native American tribes.

**witching.** A practice that attempts to locate hidden water or water-bearing structures without the use of scientific equipment; **dowsing.**

# NOTES

## Chapter 1

1. Steven I. Apfelbaum, *Nature's Second Chance* (Boston: Beacon Press, 2009). Inspired by his experiences restoring Stone Prairie Farm, Steve delves deeply into the philosophical and ethical underpinnings of ecological restoration.

2. William R. Jordan III, *The Sunflower Forest* (Berkeley and Los Angeles: University of California Press, 2003). Jordan argues that ecological restoration serves to provide a vital link between people in developed societies and the ecosystems from which they have become largely alienated; and Andre F. Clewell and James Aronson, *Ecological Restoration*, Society for Ecological Restoration International (Washington, DC: Island Press, 2007). A more advanced examination of ecological restoration with case studies from around the world.

3. Alan Haney and R. L. Power, "Adaptive Management for Sound Ecosystem Management," *Environmental Management* 20 (1996): 879–86.

## Chapter 2

1. F. H. Bormann and G. E. Likens, *Pattern and Process in a Forested Ecosystem* (New York: Springer, 1979).

2. Lucy E. Braun, *Deciduous Forests of Eastern North America* (Philadelphia: Blakiston Publishing, 1950).

3. William R. Jordan III, *The Sunflower Forest* (Berkeley and Los Angeles: University of California Press, 2003). Jordan argues that ecological restoration serves to provide a vital link between people in developed societies and the ecosystems from which they have become largely alienated; and Andre F. Clewell and James Aronson, *Ecological Restoration*, Society for Ecological Restoration International (Washington, DC: Island Press, 2007). A more advanced examination of ecological restoration with case studies from around the world.

4. Andre F. Clewell and James Aronson, "Reference Models and Developmental Trajectories," chapter 5 in *Ecological Restoration: Principles, Values and Structure of an Emerging Profession* (Washington DC: Island Press, 2007).

## Chapter 3

1. Andre F. Clewell and James Aronson, "Reference Models and Developmental Trajectories," chapter 5 in *Ecological Restoration: Principles, Values and Structure of an Emerging Profession* (Washington DC: Island Press, 2007).

2. Alan Haney and M. S. Boyce, introduction to *Ecosystem Management: Applications for Sustainable Forest and Wildlife Resources*, M. S. Boyce and Alan Haney, editors (New Haven: Yale University Press, 1997), 1–17.

## Chapter 4

1. John Madson, *Where the Sky Began: Land of the Tallgrass Prairie* (Boston: Houghton Mifflin, 1982).

2. W. S. Alverson, W. Kuhlmann, and D. M. Waller, *Wild Forests: Conservation Biology and Public Policy* (New York: Island Press, 1994).

3. S. I. Apfelbaum, F. F. Faessler, and R. Baller, 1997. "Obtaining and Processing Seeds" in *The Tallgrass Restoration Handbook*, by Stephen Packard and Cornelia F. Mutel (New York: Island Press), 99–126.

4. Alianor True, editor, *Wildfire: A Reader* (New York: Island Press, 2001).

5. S. Nielsen, C. Kirschbaum, and A. Haney, "Restoration of Midwest Oak Barrens: Structural Manipulation of Process-only?" *Conservation Ecology* 7 (2003), htpp://www.consecol.org/vol7/iss2/art10; and Alan Haney, M. Bowles, S. Apfelbaum, E. Lain, and T. Post, "Gradient Analysis of an Eastern Sand Savanna's Wood Vegetation, and Its Long-term Responses to Restored Fire Processes," *Forest Ecology and Management* 256 (2008): 1560–71.

## Chapter 5

1. Luna Leopold, *Water, Rivers, and Creeks* (Sausalito, CA: University Sciences Books, 1997); Nancy D. Gordon, Brian L. Findlayson, Thomas A. McMahon, and Christopher J. Gippel, *Stream Hydrology*, 2nd ed. (New York: John Wiley, 2004); and D. L. Brakensiek, H. B. Osborn, and W. J. Rawls, *Field Manual for Research in Agricultural Hydrology*. Agriculture Handbook 224, U.S. Department of Agriculture, http://naldr.nal.usda.gov/NALWeb/Agricola_Link.asp?Accession=CAT87209759.

2 . Robert W. Pennak, *Freshwater-invertebrates of United States*, 2nd ed. (New York: John Wiley, 1978).

## Chapter 6

1. Frank Bergon, *The Journals of Lewis and Clark* (New York: Penguin Books, 1989); and Eliza Farnham, *Life in Prairie Land* (Urbana and Chicago: University of Illinois Press, 1988). Reprint of 1846 original publication by Harper: New York.

2. J. E. Weaver and F. W. Albertson, *Grasslands of the Great Plains: Their Nature and Use* (Lincoln, NE: Johnson Publishing, 1956).

3. Wayne R. Pauly, *How to Manage Small Prairie Fires* (Madison, WI: Dane County Park Commission, 2002). To obtain a copy, send $3.00 to 4315 Robertson Road, Madison, WI 53714.

## Chapter 7

1. M. S. Boyce and Alan Haney, *Ecosystem Management: Application for Sustainable Forest and Wildlife Resources* (New Haven: Yale University Press, 1996).

2. Stephen J. Pyne, *Tending Fire* (Washington, DC: Island Press, 2004).

3. Lucy E. Braun, *Deciduous Forests of Eastern North America* (Philadelphia: Blakiston Publishing, 1950).

4. W. S. Alverson, W. Kuhlmann, and D. M. Waller, *Wild Forests: Conservation Biology and Public Policy* (Washington, DC: Island Press, 1994).

5. R. C. Anderson, J. S. Fralish, and J. M. Baskin, *Savannas, Barrens, and Rock Outcrop Plant Communities of North America* (Cambridge: Cambridge University Press, 1999).

6. Peter Friederici, editor, *Ecological Restoration of Southwestern Ponderosa Pine Forests* (Washington, DC: Island Press, 2003).

7. Steven Packard, *The Tallgrass Restoration Handbook: For Prairies, Savannas, and Woodlands* (New York: Island Press, 1997).

8. Friederici, *Ecological Restoration of Southwestern Ponderosa Pine Forests.*

9. Thomas M. Bonnicksen, *America's Ancient Forests: From the Ice Age to the Age of Discovery* (New York: John Wiley, 2000).

## Chapter 8

1. J. A. Kusler, *Our National Wetland Heritage: A Protection Guidebook* (Washington, DC: Environmental Law Institute, 1983).

2. C. H. Bartlett, *Tales of Kankakee Land* (New York: Scribner's, 1904).

3. J. H. Davis, *The Ecology and Geologic Role of Mangroves in Florida*, Publication no. 517 (Washington, DC: Carnegie Institution,1940), 303–412; M. S. Douglass, *The Everglades: River of Grass* (New York: Ballantine, 1947); H. Walter, *Vegetation of the Earth* (New York: Springer-Verlag, 1973); and M. L. Heinselman, "Landscape Evolution and Peatland Types, and the Lake Agassiz Peatlands Natural Area, Minnesota," *Ecological Monographs* 40 (1970): 235–61.

4. R. W. Tiner, *Wetlands of the United States: Current Status and Recent Trends.* National Wetlands Inventory (Washington, DC: U.S. Fish and Wildlife Service, Department of Interior, 1984).

5. A. G. Van der Valk and C. B. Davis, "The Role of Seed Banks in Vegetation Dynamics of Prairie Glacial Marshes," *Ecology* 59 (1978): 322–35.

6. S. I. Apfelbaum, "Cattail (*Typha* spp.) Management," *Natural Areas Journal* 5 (1985): 9–17.

7. Heinselman, "Landscape Evolution and Peatland Types."

8. The full text of the federal Clean Water Act, with definitions, can be viewed at www.epa.gov/npdes/pubs/cwatxt.txt.

9. See the U.S. Fish and Wildlife Service National Wetlands Inventory at www.fws.gov/wetlands/.

10. W. J. Mitsch and J. G. Gosselink, *Wetlands* (New York: Van Nostrand Reinhold, 1986).

11. S. I. Apfelbaum, J. D. Eppich, and J. Solstad, Runoff Management, Wetland Hydrology, and Biodiversity Relations in Minnesota's Red River Basin Wetlands. Unpublished manuscript.

## Chapter 9

1. G. E. Likens and F. H. Bormann, *Biogeochemistry of a Forested Ecosystem*, 2nd ed. (New York: Springer-Verlag, 1995).

2. David Rosgen, "Rosgen Geomorphic Channel Design," chapter 11, in *Stream Restoration Design* by J. Bernard, J. F. Fripp, and K. R. Robinson, editors, National Engineering Handbook, Natural Resources Conservation Service, U.S. Department of Agriculture. Washington, DC: 2007; and David L. Rosgen, *River Stability Field Guide* (Fort Collins, CO: Wildland Hydrology Consultants, 2008).

3. Luna Leopold, *Water, Rivers, and Creeks* (Sausalito, CA: University Sciences Books, 1997); Nancy D. Gordon, Brian L. Findlayson, Thomas A. McMahon, and Christopher J. Gippel, *Stream Hydrology*, 2nd ed. (New York: John Wiley, 2004); and D. L. Brakensiek, H. B. Osborn, and W. J. Rawls, *Field Manual for Research in Agricultural Hydrology*. Agriculture Handbook 224, U.S. Department of Agriculture, http://naldr.nal.usda.gov/NALWeb/Agricola_Link.asp?Accession=CAT87209759.

4. J. Bernard, J. F. Fripp, and K. R. Robinson, editors, *Stream Restoration Design*, chapter 14 of the National Engineering Handbook. Natural Resources Conservation Service, U.S. Department of Agriculture, Washington, DC: 2007. (This can be downloaded free from www.toodoc.com/National-Engineering-Handbook-ebook.html).

## Chapter 10

1. Victor E. Shelford, *Naturalist's Guide to the Americas* (Baltimore: Williams and Wilkins Company, 1926).

2. F. Shreve, "The Desert Vegetation of North America," *Botanical Review* 8 (1942): 195–247.

3. Ibid.

4. David Bainbridge, *Guide for Desert and Dry Land Restoration: New Hope for Arid Lands* (Covelo, CA: Island Press, 2007).

5. Ibid.

6. J. E. Lovich, *Human Induced Changes in the Mojave and Colorado Desert Ecosystems: Recovery and Restoration Potential* (Riverside, CA: U.S. Geological Survey, Biological Resources Division, University of California, 2002).

7. D. H. Gray and A. T. Leiser, *Biotechnical Slope Protection and Erosion Control* (Malabar, FL: Van Nostrand Reinhold, 1989); and California Plant Health Association, *Western Fertilizer Handbook,* 9th ed. (Danville, IL: Prentice-Hall, 1989).

8. L. L. St. Clair, B. L. Webb, J. R. Johansen, and G. T. Nebeker, "Cryptogamic Soil Crusts: Enhancement of Seedling Establishment in Disturbed and Undisturbed Areas," *Reclamation and Revegetation Research* 3 (1984): 129–36.

9. Bainbridge, *Guide for Desert and Dry Land Restoration.*

10. California Plant Health Association. *Western Fertilizer Handbook.*

# INDEX

Page references followed by "f" refer to text figures.

# Island Press | Board of Directors


Alexis G. Sant *(Chair)*
Managing Director
Persimmon Tree Capital

Katie Dolan *(Vice-Chair)*
Executive Director
The Nature Conservancy
of Eastern NY

Henry Reath *(Treasurer)*
Nesbit-Reath Consulting

Carolyn Peachey *(Secretary)*
President
Campbell, Peachey & Associates

---

Decker Anstrom
Board of Directors
Comcast Corporation

Stephen Badger
Board Member
Mars, Inc.

Katie Dolan
Eastern New York
Chapter Director
The Nature Conservancy

Merloyd Ludington Lawrence
Merloyd Lawrence, Inc.
and Perseus Books

William H. Meadows
President
The Wilderness Society

Pamela B. Murphy

Drummond Pike
President
The Tides Foundation

Charles C. Savitt
President
Island Press

Susan E. Sechler

Victor M. Sher, Esq.
Principal
Sher Leff LLP

Peter R. Stein
General Partner
LTC Conservation Advisory
Services
The Lyme Timber Company

Diana Wall, Ph.D.
Director, School of Global
Environmental Sustainability
and Professor of Biology
Colorado State University

Wren Wirth
President
Winslow Foundation